Atoms and Energy

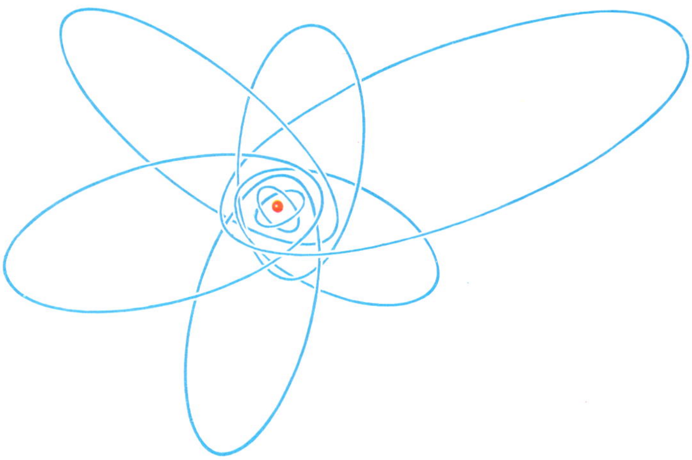

WARWICK PRESS

Contents

Author Neil Ardley
Editor John Paton

Library of Congress Cataloging in Publication Data

Ardley, Neil.
Atoms and energy.

(Modern knowledge library)
Includes index.

SUMMARY: Discusses the structure of the atom and its role in the production of nuclear energy.

1. Atomic energy—Juvenile literature. 2. Atoms—Juvenile literature. [1. Atomic energy. 2. Atoms] I. Title.

QC778.5.A7 1976 539.7 76-13597
ISBN 0-531-02442-3
ISBN 0-531-01200-X lib. bdg.

First published in Great Britain by Sampson Low in 1975
Printed in Italy by New Interlitho, Milan

6 5 4 3 2 1

Atoms and Energy

Everything you see – trees, water, paper, steel – is made up of smaller things called *atoms*. The ancient Greeks were the first people to think about atoms. They wondered what would happen if a piece of wood was cut into smaller and smaller pieces. Sooner or later, they reasoned, each part would be so small it couldn't be seen or be divided any more.

We now know the Greeks were not quite right. Even an atom can be divided into smaller parts. In fact, by splitting atoms scientists have made atomic explosions and nuclear power. The tiny atom can be the source of a lot of energy.

It may surprise you but there are fewer than a hundred different kinds of atoms. However, these atoms combine into families called *molecules*, and there are millions of different molecules. A molecule of water, for example, is made of two hydrogen atoms and one oxygen atom. A molecule of hydrogen peroxide is made up of two hydrogen atoms and two oxygen atoms.

The way atoms combine determines the look and nature of everything around us. Solids, liquids and gases are all made up of atoms. But the atoms in a solid are packed together more tightly than they are in a liquid or a gas. The most amazing thing is that a table (which looks and feels solid) is mostly empty space. The molecules that make up the table are very far apart. The atoms which make up the molecules are also far apart.

Solid

Liquid

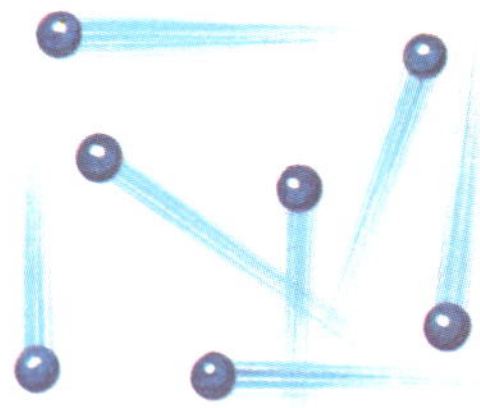

Gas

What are Atoms Like?

Atoms are unbelievably small. Yet everything is made of them from the smallest plant to the largest animal, from a grain of sand to a world in space.

It is hard to understand something as small as an atom. If we say that 250,000 million of them make up the period at the end of this sentence it really means very little to us. So it helps to think of an atom as being the size of something you know. Then compare it to another familiar object. For example, if the smallest atom were the size of the nail on your little finger, then your hand would be as big as the world! That gives you a rough idea of the size of the smallest atom.

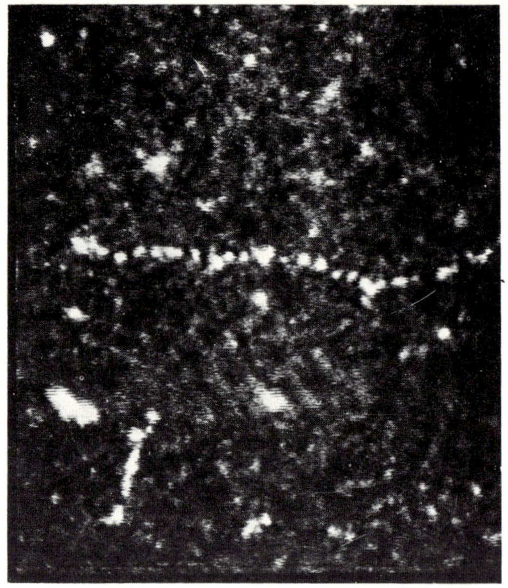

Right: A photograph of a chain of thorium atoms. (Thorium is a soft, radioactive metal). Each white dot is one atom. The photograph was taken with an electron microscope at the University of Chicago.

The Arrangement of Atoms

We do not really know what an atom looks like. It is possible to take pictures of large atoms through special microscopes, but the pictures show only white blobs. We do know, however, that the atoms in an object do not look like small pieces of the object. A hydrogen atom from red paint looks the same as one from colorless water. The appearance of anything depends on the way the atoms are arranged.

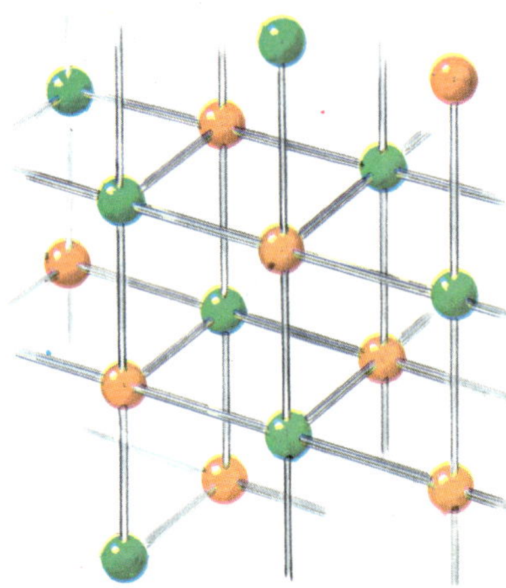

Right: How atoms join together in common salt. Sodium and chlorine ions come together to form a lattice. Repeated indefinitely, such a lattice becomes a crystal of salt.

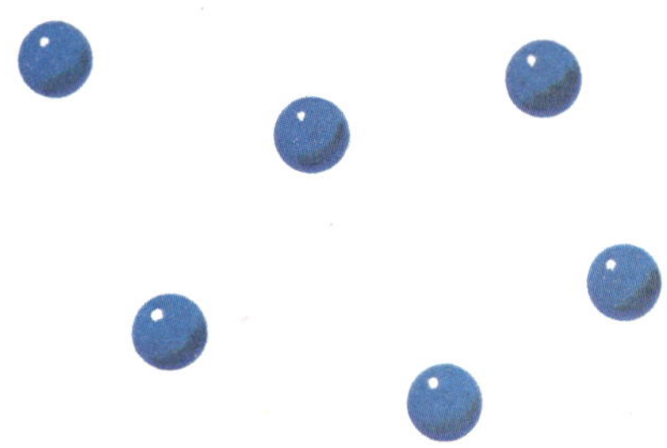

Right: Water molecules are completely free and can move in all directions.

Left: Salt dissolving in water. The sodium and chlorine ions move away from the lattice and water molecules burst it apart.

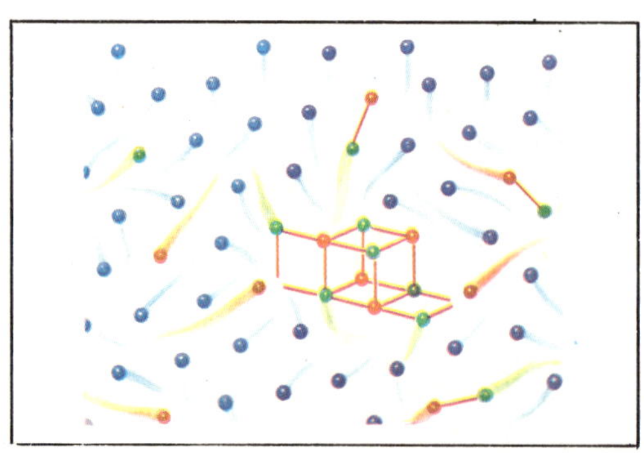

Right: Salt condensing into crystals. As the salt solution becomes more condensed, sodium and chlorine ions join together in pairs. Ultimately they form a crystal lattice.

The arrangement of atoms or molecules in anything also determines whether it is a gas like the air, liquid like water, or solid like the chair on which you are sitting. There are forces between the atoms or molecules that pull them together. In a solid, the forces are strong and the atoms or molecules pack tightly together. In a liquid, the forces are not so strong. This is why a liquid can flow to take up any shape and other objects can move through it. In gases, the forces are very weak and the gas atoms or molecules are not arranged in any particular way. They can move apart until the gas fills its container.

Heat Makes Atoms Vibrate

We can change a solid into a liquid and a liquid into a gas by heating it. The solid melts and the liquid boils. This happens because heat reduces the forces linking the atoms or molecules. The atoms or molecules are always vibrating (moving), even in the tight arrangement of a solid. The hotter they get, the more they vibrate. If a solid is heated enough, the atoms or molecules vibrate so much that the arrangement becomes loose and the solid melts to liquid. Add more heat and the atoms or molecules move even more and finally break away from the arrangement. The liquid boils and becomes a gas. In a gas, the atoms or molecules are flying about in all directions.

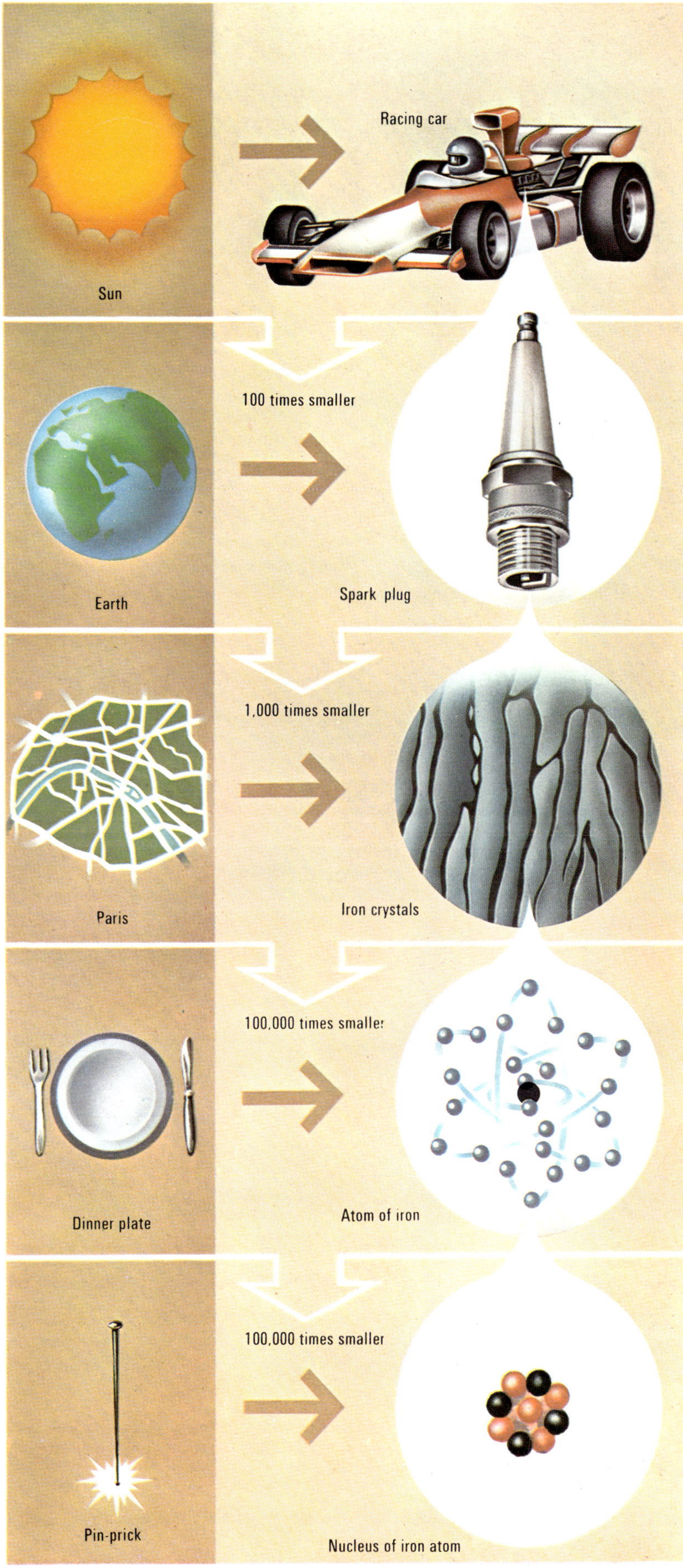

Right: The atom is unbelievably small. It is so small that if a car were enlarged to the size of the sun, one of its spark plugs would be the size of the earth; a molecule in the metal of the spark plug would be as big as a city the size of Paris; an atom in the molecule would be the size of a dinner plate, and the nucleus of the atom (its center) would be as small as a pin-prick.

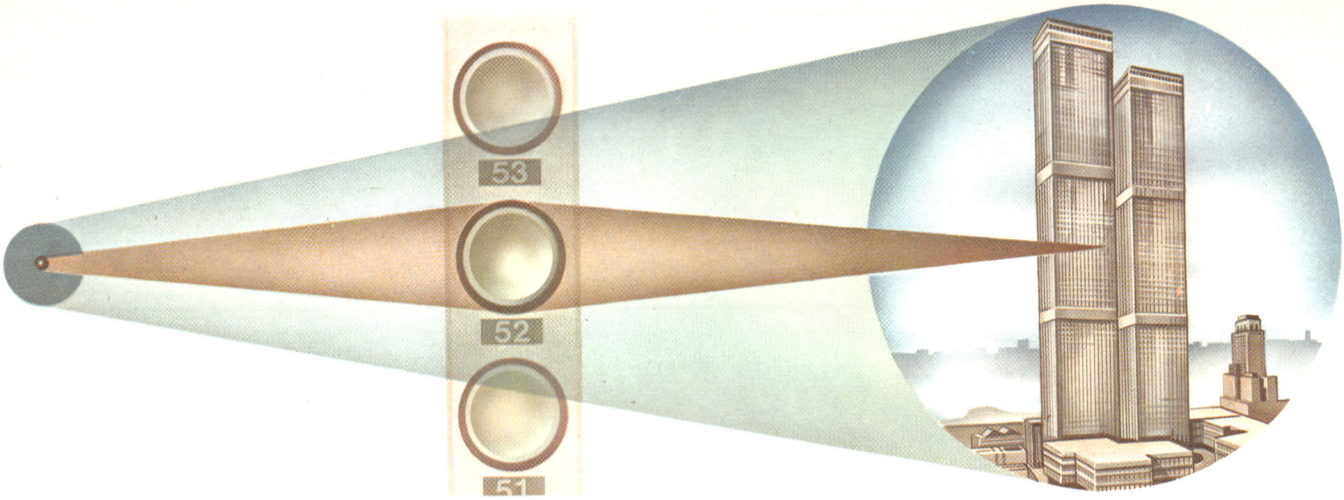

What are Atoms Made of?

An atom is made up of a tiny central nucleus surrounded by electrons. If the nucleus were the size of an elevator button the whole atom would be the size of a skyscraper.

Despite its size, the atom is like a miniature solar system. Most of it is empty space.

If you could see an atom in detail, there would really be very little to see. The outside of the atom consists of tiny particles called *electrons*. The electrons each have an electric charge. They move so fast around the center (nucleus) of the atom that the outside is like a cloud of electric charge. The electrons move in different paths or orbits. When one jumps to an orbit closer to the center it gives out light. An atom therefore would look like a glowing cloudy sphere.

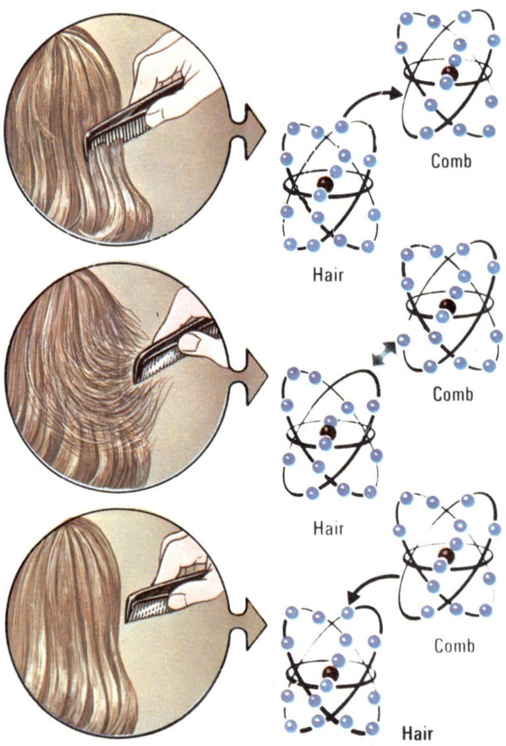

Comb your hair strongly. Some of the hairs will rise and stick to the comb. Electrons have been rubbed from the atoms in your hair to the atoms in the comb. The comb gains a negative charge and the hair a positive charge. Because unlike charges attract, the charged hairs are attracted to the comb. The hair then regains its lost electrons, and drops back.

Lightning (left) is caused by the movement of electrons from cloud to cloud, or cloud to earth.

Sir Joseph John Thomson (1856–1940) was a British physicist who discovered the electron. He found that cathode rays, which are beams of negative electricity, can be bent by an electric field. Thomson showed that the rays are made up of tiny particles. He thought that these particles were contained in all atoms. Later they were called electrons.

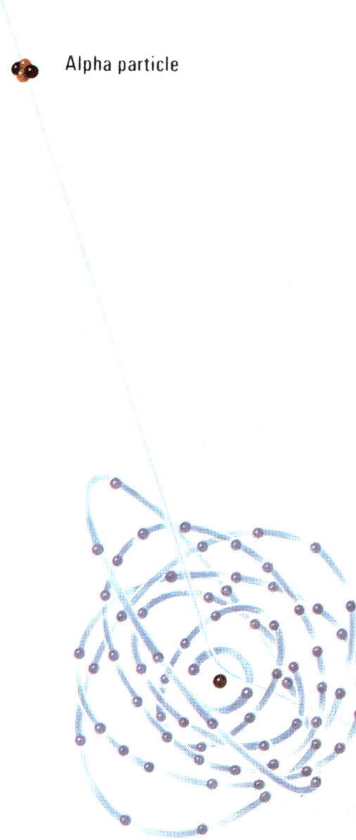

Ernest Rutherford (1871–1937) was born in New Zealand and studied under J. J. Thomson. In 1911, he became the first man to explain the structure of the atom. In earlier experiments he fired alpha particles at a film of gold. Alpha particles have a positive charge (in fact, they each consist of two protons and two neutrons). They easily penetrated the negative electrons in the gold atoms. But they were pushed aside by the nuclei in the atoms, which have a positive charge and therefore repelled the alpha particles. This showed that atoms have a small nucleus containing positively charged protons, surrounded by negatively charged electrons.

For his brilliant work on the atom Rutherford won the Nobel Prize in chemistry.

If you could look inside the atom, you would spot a tiny, solid body in the center. This body – the *nucleus* – is only one ten-thousandth the size of the atom. Between it and the electrons is empty space. Although the nucleus is very small, it is several thousand times heavier than the electrons that move around it.

The electric charge on each electron is called a *negative* charge. The nucleus has *positive* electric charges. Normally, the number of negative charges on the electrons equals the number of positive charges on the nucleus. The positive and negative charges cancel each other out, so the atom as a whole normally has no overall electric charge.

It is a law of electricity that like charges repel each other and unlike charges attract. If two atoms closely approach each other, the electrons at the outside of each atom repel each other because they have the same negative charge. The closer they approach the more strongly they push each other apart. It is impossible for one atom to move inside another. Forces other than electric forces keep the atoms together in solids and liquids.

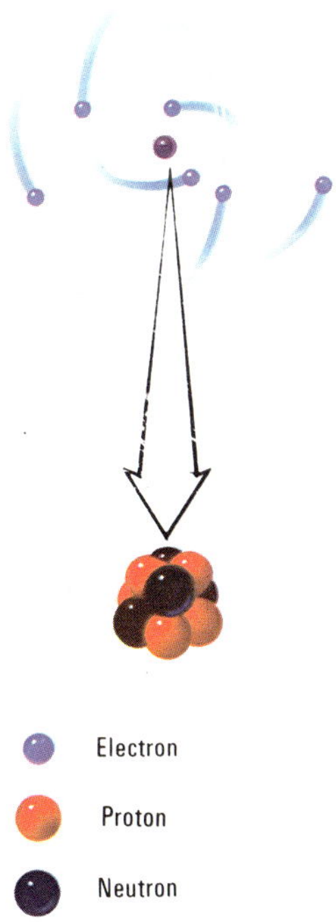

Above: A carbon atom is made up of six electrons orbiting a nucleus of six protons and six neutrons.

The nucleus of an atom is in fact not completely solid. It is made up of smaller particles grouped tightly together. One of these particles is called the *proton*. It has one positive electric charge, and is nearly 2000 times heavier than an electron. The total number of protons in the nucleus normally equals the number of electrons. This is why the positive and negative charges in an atom are usually equal in number. The simplest atoms have a nucleus of one proton surrounded by one orbiting electron. These atoms are hydrogen atoms.

All the other elements apart from hydrogen contain another kind of particle in the nucleus of their atoms. This particle is called the *neutron*. It is almost exactly like the proton, except that it has no electric charge. An atom of carbon, for example, has six electrons and therefore its nucleus contains six protons. But the nucleus also has six neutrons.

Kinds of Atoms

There are only about a hundred different kinds of atoms, yet they combine together to make up everything in the universe.

The centers of atoms are made up of protons and neutrons, around which electrons move. One atom is therefore going to be different from another if it contains different numbers of these particles. Substances that contain only one kind of atom are known as *elements*. Hydrogen, oxygen, carbon and chlorine are elements. So are metals such as mercury, copper, sodium, gold, silver and uranium. An atom of one element is different from an atom of another element because it has a different number of protons in its nucleus. A hydrogen atom has one proton, a carbon atom six and an oxygen atom eight, for example. Altogether 105 different elements are known. But some of these can only be made artificially, and

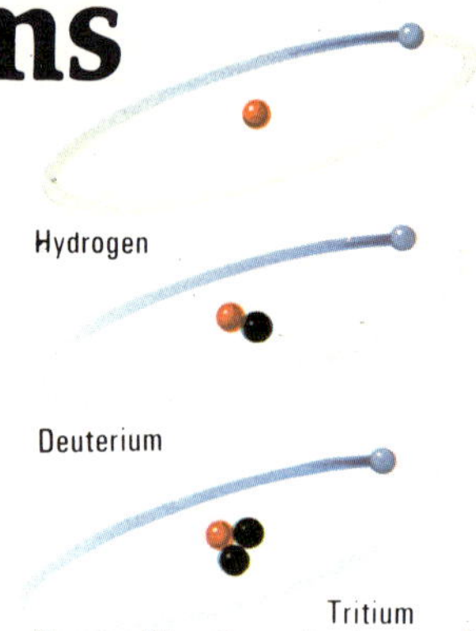

Above: The three isotopes of hydrogen. Ordinary hydrogen (hydrogen-1) has a nucleus of one proton. Hydrogen-2 or deuterium has a nucleus of one proton and one neutron. Tritium or hydrogen-3 has a nucleus of one proton and two neutrons. Only one electron orbits each nucleus.

Below: Atoms of helium-3 and helium-4 have a nucleus of two protons and one or two neutrons.

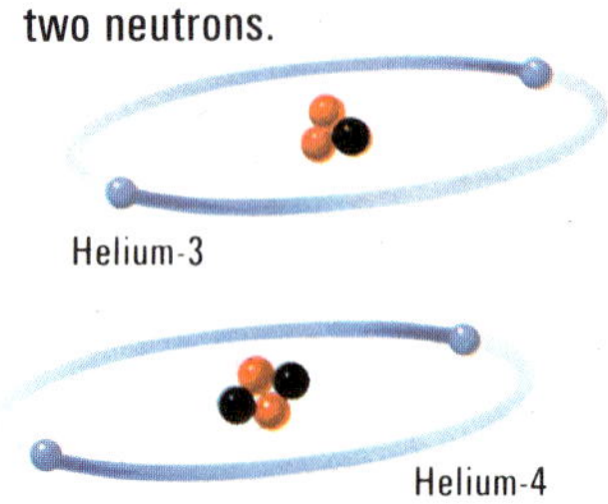

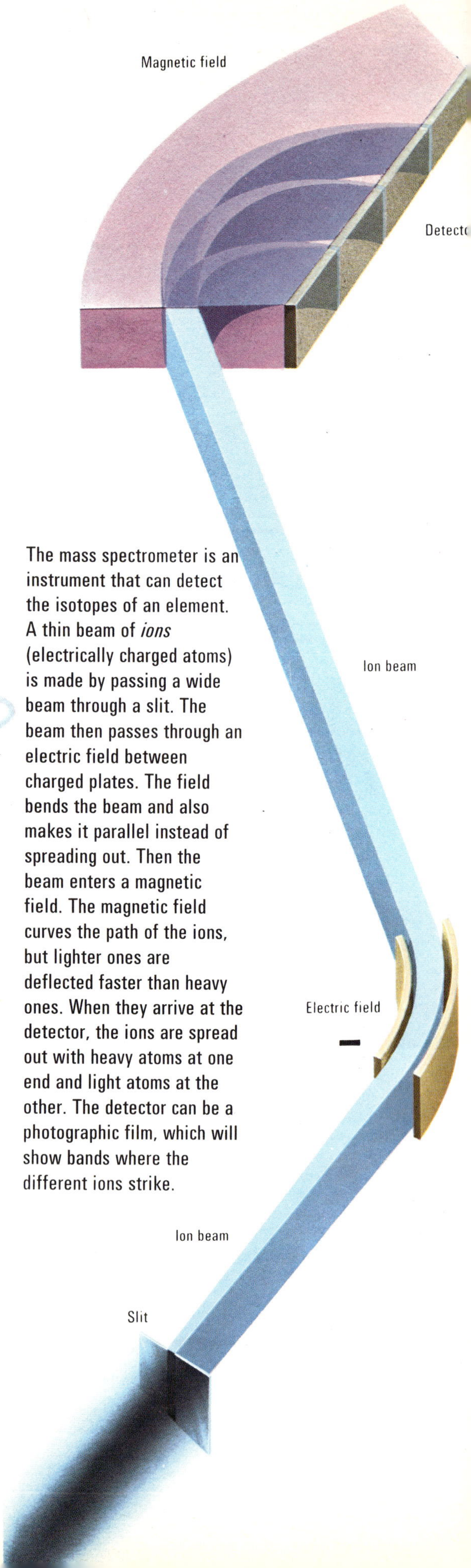

The mass spectrometer is an instrument that can detect the isotopes of an element. A thin beam of *ions* (electrically charged atoms) is made by passing a wide beam through a slit. The beam then passes through an electric field between charged plates. The field bends the beam and also makes it parallel instead of spreading out. Then the beam enters a magnetic field. The magnetic field curves the path of the ions, but lighter ones are deflected faster than heavy ones. When they arrive at the detector, the ions are spread out with heavy atoms at one end and light atoms at the other. The detector can be a photographic film, which will show bands where the different ions strike.

have a very short life. Less than 100 elements are found in nature, and the whole universe is made up of them.

But the universe obviously contains many thousands of different substances, and not only a hundred or so elements. This is because most substances consist of combinations of two or more elements. They are called *compounds*. As there are thousands of possible combinations of elements, there are thousands of different compounds.

Atoms may gain or lose electrons. This does not make much difference to an element. It is the number of protons in the nucleus that matters, and this number is called the *atomic number*. Each element has its own atomic number. Hydrogen is 1, carbon 6, oxygen 8, and so on. If an atom were to lose or gain protons, it would become an atom of another element because its atomic number would change.

What happens if the number of neutrons in the nucleus changes? In fact, very little happens. The atom is exactly the same except that it is a little lighter or heavier. Many elements are mixtures of atoms containing different numbers of neutrons. Chlorine, for example, is a mixture of atoms containing 18 or 20 neutrons per nucleus. Each chlorine nucleus has 17 protons, so that nuclei of chlorine contain 35 or 37 particles.

Atoms of an element that contain different numbers of neutrons are called *isotopes*. Chlorine is made up of two isotopes. They are called chlorine-35 and chlorine-37.

Some elements occur in nature as only one isotope. Natural aluminum, for example, is made up of only aluminum-27. Other elements are mixtures of two or more isotopes. Tin, for example, occurs as a mixture of ten isotopes.

The first isotopes were discovered by J. J. Thomson in 1911. He detected the two isotopes of the gas neon by using a mass spectrometer to separate them.

THE ELEMENTS
and their Atomic Numbers

Atomic number	Name	Symbol	Atomic number	Name	Symbol
1	Hydrogen	H	54	Xenon	Xe
2	Helium	He	55	Cesium	Cs
3	Lithium	Li	56	Barium	Ba
4	Beryllium	Be	57	Lanthanum	La
5	Boron	B	58	Cerium	Ce
6	Carbon	C	59	Praseodymium	Pr
7	Nitrogen	N	60	Neodymium	Nd
8	Oxygen	O	61	Promethium	Pm
9	Fluorine	F	62	Samarium	Sm
10	Neon	Ne	63	Europium	Eu
11	Sodium	Na	64	Gadolinium	Gd
12	Magnesium	Mg	65	Terbium	Tb
13	Aluminum	Al	66	Dysprosium	Dy
14	Silicon	Si	67	Holmium	Ho
15	Phosphorus	P	68	Erbium	Er
16	Sulfur	S	69	Thulium	Tm
17	Chlorine	Cl	70	Ytterbium	Yb
18	Argon	Ar	71	Lutetium	Lu
19	Potassium	K	72	Hafnium	Hf
20	Calcium	Ca	73	Tantalum	Ta
21	Scandium	Sc	74	Tungsten	W
22	Titanium	Ti	75	Rhenium	Re
23	Vanadium	V	76	Osmium	Os
24	Chromium	Cr	77	Iridium	Ir
25	Manganese	Mn	78	Platinum	Pt
26	Iron	Fe	79	Gold	Au
27	Cobalt	Co	80	Mercury	Hg
28	Nickel	Ni	81	Thallium	Tl
29	Copper	Cu	82	Lead	Pb
30	Zinc	Zn	83	Bismuth	Bi
31	Gallium	Ga	84	Polonium	Po
32	Germanium	Ge	85	Astatine	At
33	Arsenic	As	86	Radon	Rn
34	Selenium	Se	87	Francium	Fr
35	Bromine	Br	88	Radium	Ra
36	Krypton	Kr			
37	Rubidium	Rb	89	Actinium	Ac
38	Strontium	Sr	90	Thorium	Th
39	Yttrium	Y	91	Protactinium	Pa
40	Zirconium	Zr	92	Uranium	U
41	Niobium	Nb	93	Neptunium	Np
42	Molybdenum	Mo	94	Plutonium	Pu
43	Technetium	Tc	95	Americium	Am
44	Ruthenium	Ru	96	Curium	Cm
45	Rhodium	Rh	97	Berkelium	Bk
46	Palladium	Pd	98	Californium	Cf
47	Silver	Ag	99	Einsteinium	Es
48	Cadmium	Cd	100	Fermium	Fm
49	Indium	In	101	Mendelevium	Mv
50	Tin	Sn	102	Nobelium	No
51	Antimony	Sb	103	Lawrencium	Lw
52	Tellurium	Te	104	Rutherfordium	Rf
53	Iodine	I	105	Hahnium	Ha

EVERYTHING ON THE EARTH IS MADE OF ATOMS OF THESE ELEMENTS OR COMBINATIONS OF THEM

How do Atoms Join Together?

Atoms may join together in simple patterns or in long chains.

Almost all the substances we use in our everyday lives – such as water, salt, wood, cloth, food and medicines – are made of compounds rather than elements. To make compounds, atoms of elements must join together. This can happen in two ways.

Atoms can gain or lose electrons to form *ions*. If an atom gains an electron, it also gains the negative electric charge of the electron. It becomes a negative ion. If an atom loses an electron, it loses the negative charge and becomes a positive ion. Because unlike electric charges attract each other, a negative ion and positive ion will be pulled together, rather like the way a magnet attracts iron. A bond called an *ionic bond* forms between the two atoms. Many atoms can join together in this way to make up a tight group of atoms.

The other way in which atoms can join together is by sharing electrons. When an ionic bond forms, one atom gives one or more electrons to another atom. The electrons move from one atom to the other. But in many compounds, electrons are given by both the atoms that join. The electrons move around both atoms and form a bond called a *covalent bond*. In some kinds of covalent bonds, the electrons come from one of the joining atoms instead of from both.

When ionic bonds form, the electrical attraction can bring any number of atoms together in a great arrangement of atoms. The arrangement has no set size, and more atoms can simply join on to it if necessary. But when covalent bonds form between atoms, a definite group of

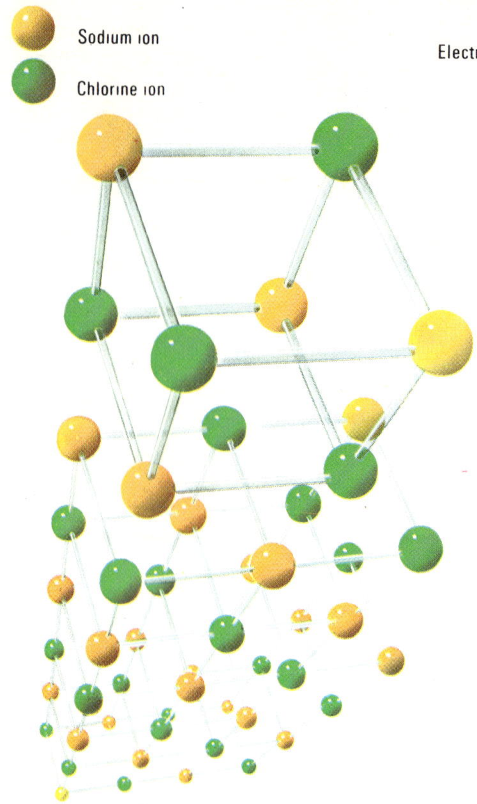

Above: Common salt is made up of sodium and chlorine ions joined together by ionic bonds in a cube-shaped network.

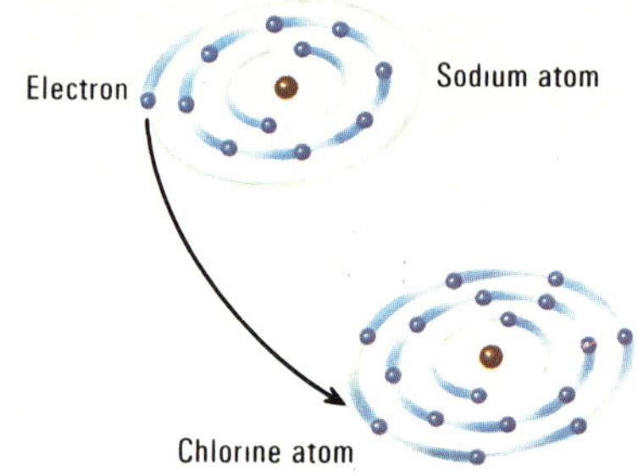

Salt is an ionic compound made up of sodium and chlorine ions. When the compound forms, each sodium atom loses one electron to a chlorine atom (above). The positive sodium ion and negative chlorine ion are attracted to each other (below).

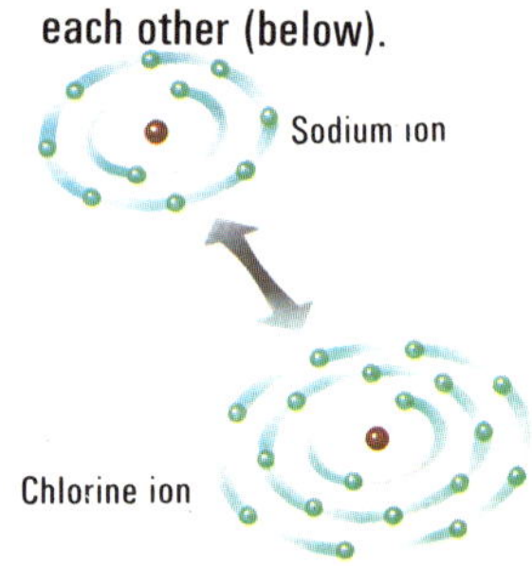

Above: Water consists of many molecules. Each one is made up of two hydrogen atoms and one oxygen atom joined together.

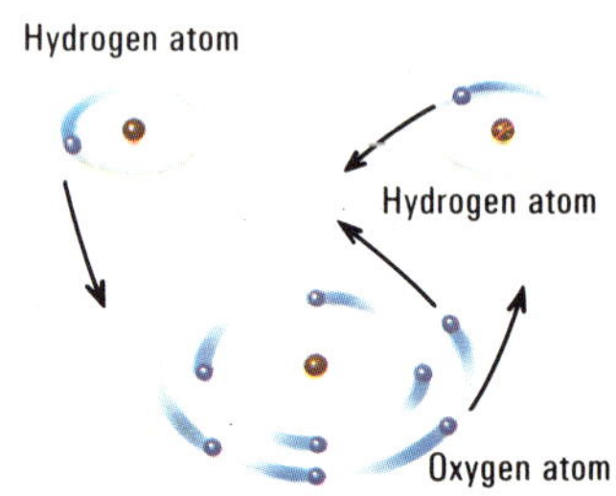

Water is a covalent compound made up of hydrogen and oxygen. Two hydrogen atoms join with one oxygen atom. Each hydrogen atom gives one electron to form the bond and the oxygen atom gives two electrons (above). Two electrons move around the two atoms, holding the atoms together (below). The three atoms join to make a water molecule.

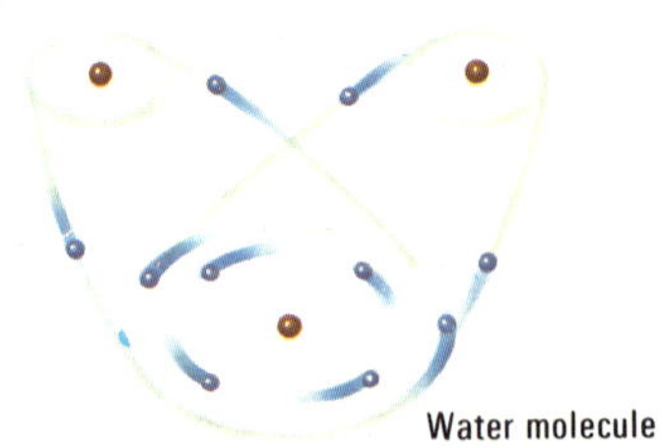

atoms join together. If there are more atoms around, they form separate identical groups. These groups are called *molecules.*

The way in which atoms join to make a compound may give the compound a certain shape. Salt, for example, is always found as cube-shaped crystals. The salt crystals in a packet of salt are very small, but it is possible to grow large cube-shaped crystals of salt. The crystals have a cubic shape because the atoms in them are joined together in a cube-shaped network.

But we cannot always grow crystals of substances to find out how the atoms are arranged inside them. Instead scientists pass a beam of X-rays through a substance. The beam is split up by the atoms in the crystal and makes a pattern on a photographic film. From the pattern, scientists can tell how the atoms are arranged in the crystal.

Above: Crystals of calcite.
Below: An X-ray crystal pattern.

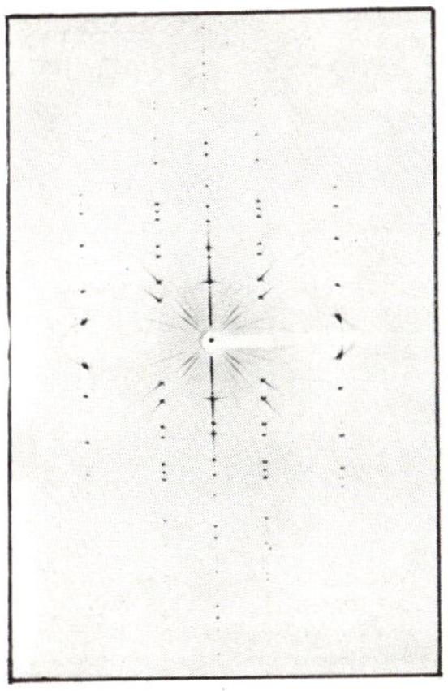

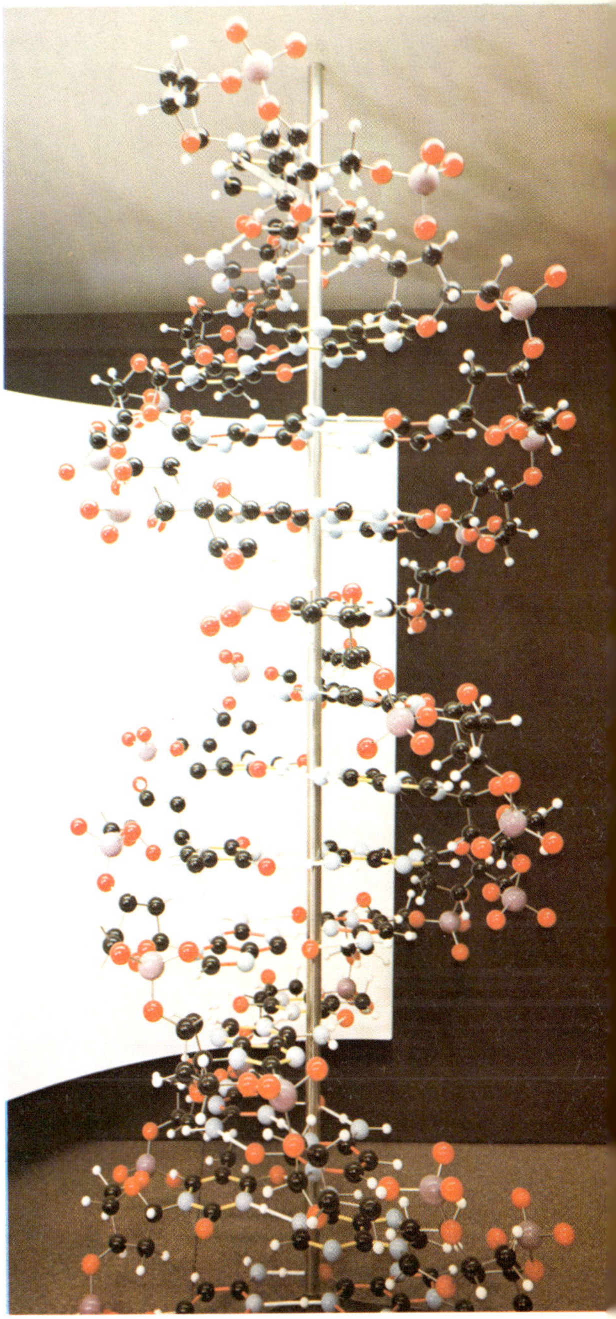

Right: A model showing the arrangement of atoms in a long molecule of DNA. It is the DNA in living cells that controls the heredity of all animals and plants.
Below: A molecule of DNA, as seen by the electron microscope (central line). The magnification is 60,000.

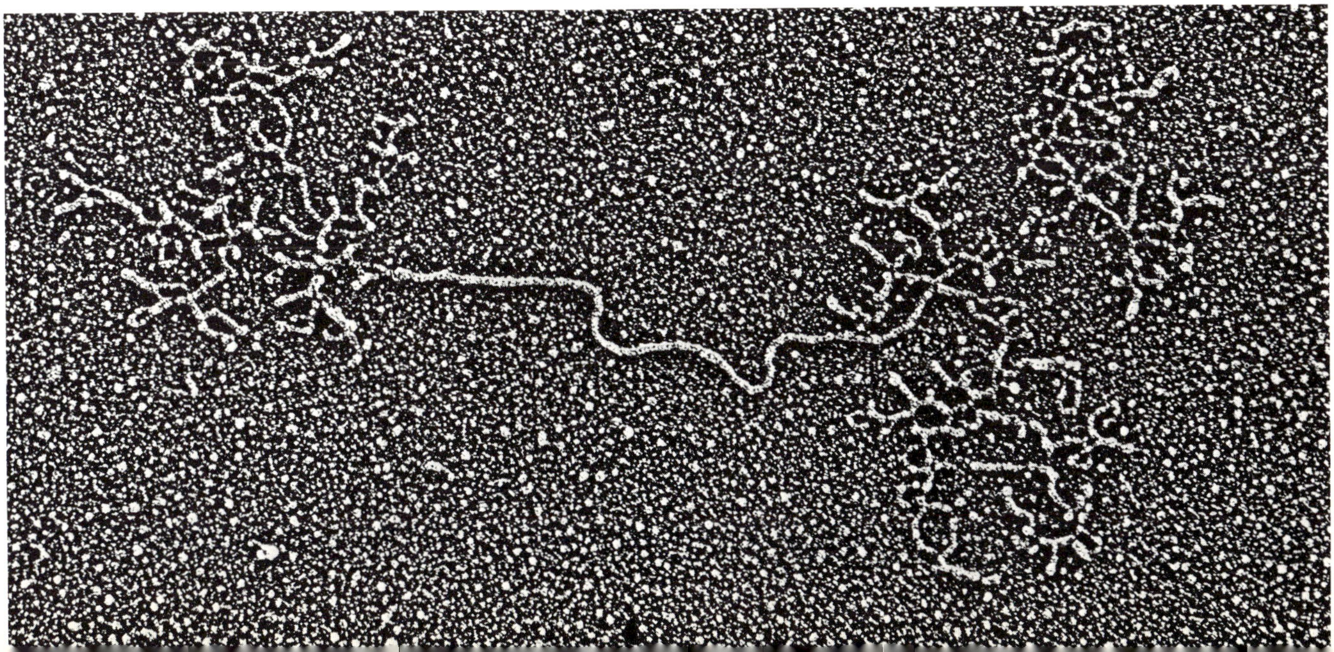

How do Atoms Break Apart?

When atoms break apart, a dream of the ancient alchemists comes true – metals change into other metals.

Alpha Decay

Some elements are not found in nature because their atoms break apart or decay. This can happen in three ways. Alpha decay is one of them. The nucleus of the atom suddenly splits and produces an alpha particle. This particle is made up of two protons and two neutrons. Alpha decay happens naturally to some elements and not others. These elements produce streams of alpha particles as their atoms break apart. The streams of particles are known as alpha rays. Alpha is the first letter of the Greek alphabet and is written α.

Above: When an atom breaks and produces an alpha particle, it loses two protons and two neutrons from its nucleus. Radium-226 (atomic number 88) becomes radon-222 (atomic number 86) by alpha decay.

Left: When beta decay happens, the nucleus loses an electron. A neutron is like a proton plus an electron, so losing an electron is like changing a neutron into a proton. The element changes to another element with an atomic number greater by 1. Iodine-128 (53 protons) becomes xenon-128 (54 protons) by beta decay.

Beta Decay

Some other elements decay in another way, known as beta decay. The nucleus suddenly produces an electron. This electron is also known as a beta particle, and streams of these electrons are called beta rays. Like alpha decay, it occurs naturally in some elements and cannot be stopped. Beta is the second letter of the Greek alphabet and is written β.

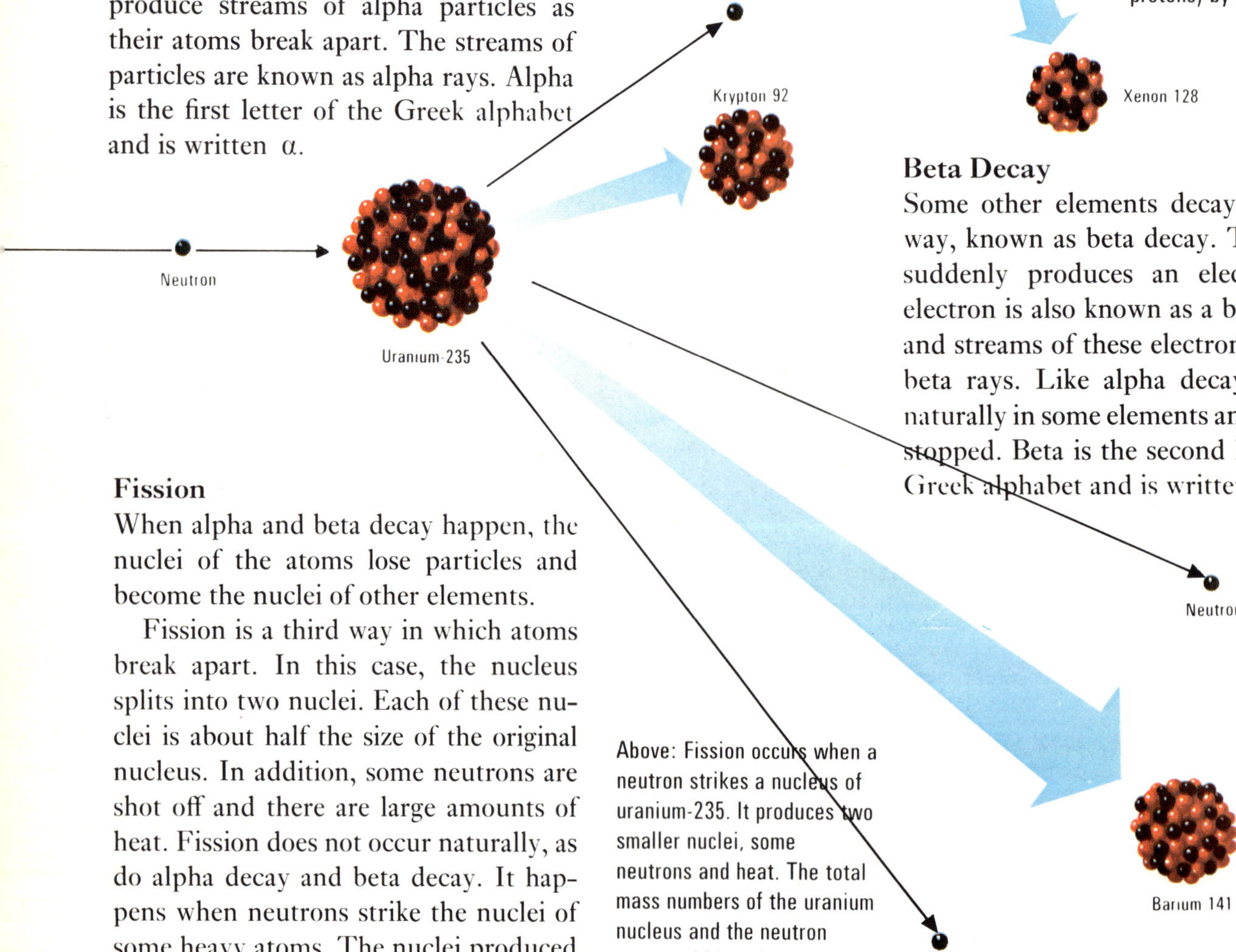

Fission

When alpha and beta decay happen, the nuclei of the atoms lose particles and become the nuclei of other elements.

Fission is a third way in which atoms break apart. In this case, the nucleus splits into two nuclei. Each of these nuclei is about half the size of the original nucleus. In addition, some neutrons are shot off and there are large amounts of heat. Fission does not occur naturally, as do alpha decay and beta decay. It happens when neutrons strike the nuclei of some heavy atoms. The nuclei produced are called fission fragments.

Above: Fission occurs when a neutron strikes a nucleus of uranium-235. It produces two smaller nuclei, some neutrons and heat. The total mass numbers of the uranium nucleus and the neutron come to 236, and so must the products.

Henri Becquerel

Henri Becquerel (1852–1908) was a French physicist who discovered radioactivity in 1896. He found that uranium compounds would fog a photographic plate, even though it had not been exposed to light. These rays were in fact beta rays. Becquerel showed that they were streams of electrons.

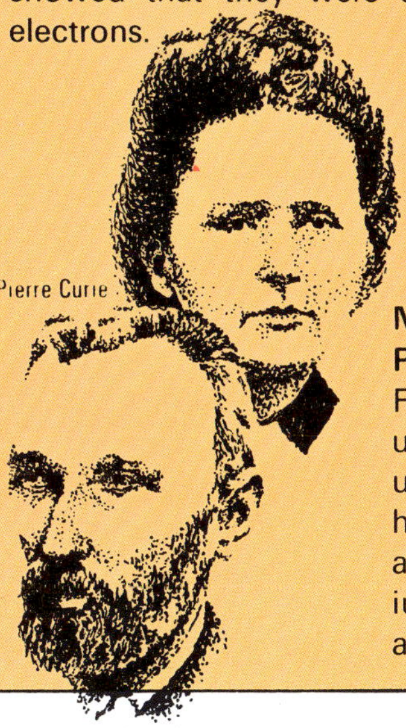

Marie Curie

Pierre Curie

Marie Curie (1867–1934) and **Pierre Curie** (1859–1906) were French chemists. They looked for an unknown source of radioactivity in uranium minerals. After many years of hard work, they obtained a small amount of radium from tons of uranium ore. It was extremely radioactive, and produced intense alpha rays.

Changing Common Metals into Gold

One of the greatest desires of the alchemists of the Middle Ages was to change common metals into gold. They tried all sorts of potions, but the metals refused to alter themselves in any way. Men did not achieve transmutation (the changing of one element into another) until 1919. A way was found to change the nucleus of the atom. The alchemists, although they knew a great deal about chemistry, knew nothing of the structure of matter.

But transmutation was going on all around the alchemist by alpha and beta decay which is called *radioactivity*. Nowadays we can add to these natural transmutations by breaking apart and building up nuclei in powerful machines called particle accelerators and in nuclear reactors. This branch of science is called *nuclear physics*.

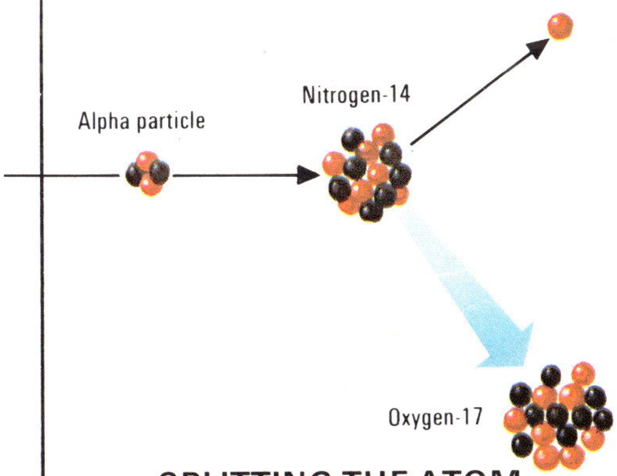

SPLITTING THE ATOM

In 1919, Ernest Rutherford became the first man to change one element into another. He bombarded nitrogen-14 with alpha particles. He found that protons were produced by the collisions of the alpha particles and nitrogen nuclei. Oxygen-17 nuclei were also produced. Rutherford had therefore changed nitrogen into oxygen by bombarding it with alpha particles. Although this feat is often called "splitting the atom", it is not a good name. Rutherford was building up atoms rather than splitting them.

An alchemist with his assistants at work in his laboratory. The attempt to change common metals into gold was known as the "Great Work".

Radiation and its Dangers

We can live with natural radiation, but we must protect ourselves from the dangerous man-made kind.

Left: The international radiation symbol. Radioactive areas and containers are all marked with the symbol.

Below: The three kinds of radiation have different penetrating powers. An alpha particle can be stopped by a piece of paper. A beta particle is more penetrating but can be stopped by a thin sheet of aluminum. A gamma ray is very penetrating. It can even pass through a piece of iron a foot long. But it gets weaker the farther it goes.

When alpha particles strike atoms, they change the nuclei of the atoms. They turn them into atoms of other elements. If beta particles, which are electrons, strike atoms, they will give the atoms an electric charge.

If alpha and beta rays strike the human body, the particles in them change the atoms there. This can be very dangerous, because our bodies are made up of many chemical compounds working together in a delicate balance. If the compounds in our bodies suddenly start changing, illness, perhaps even death, may result.

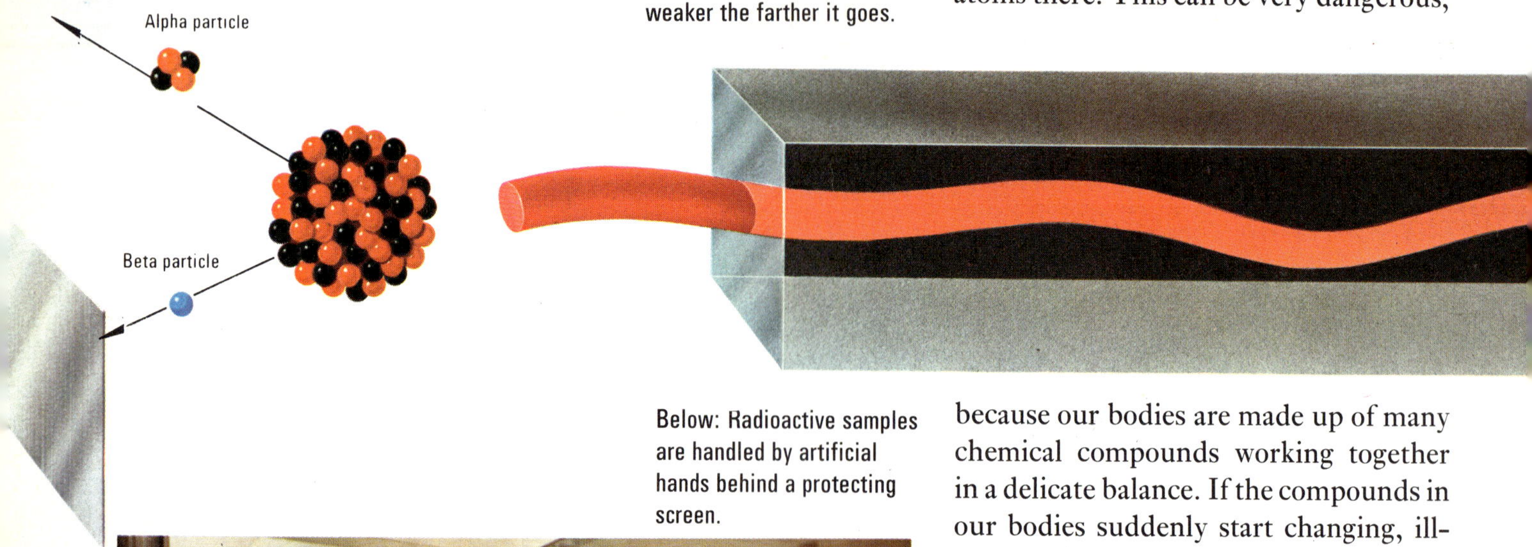

Below: Radioactive samples are handled by artificial hands behind a protecting screen.

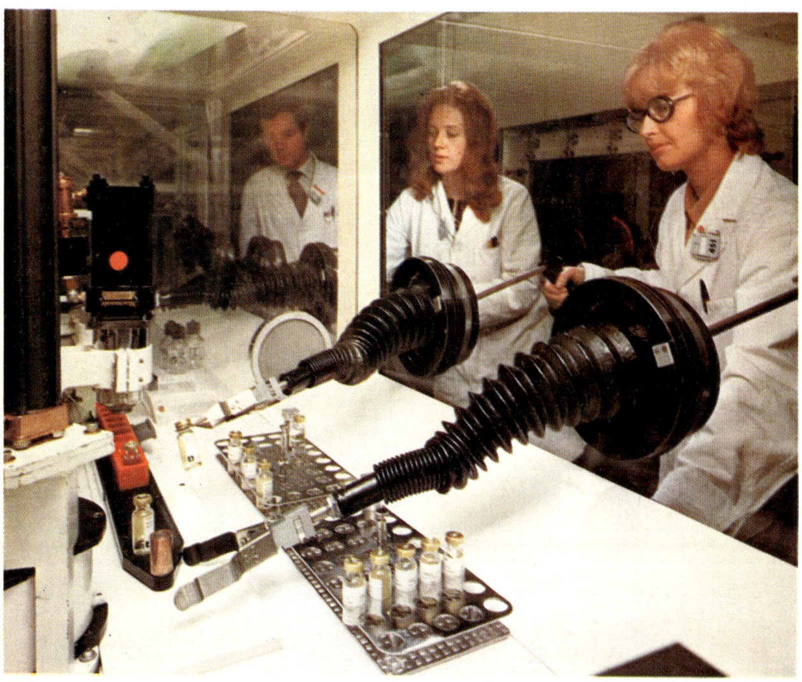

Dangerous rays and particles are often called *radiation*. Alpha and beta rays are two kinds of radiation. There is a third kind of radiation, called *gamma rays*. (Gamma is the third letter of the Greek alphabet and written γ.) Gamma rays are not streams of tiny particles like alpha and beta rays. They are a form of very strong X-rays and, like X-rays, gamma rays can easily go right through things.

Alpha particles are not very dangerous to us. They are in fact stopped by the outer layer of the skin. Beta rays can get inside the skin. But gamma rays can pass right through the body, just as X-rays can. They cause harm to the atoms in the body. X-rays also cause harm. This is

Right: A man enters a radioactive area in a nuclear energy station. He wears a thin plastic suit to protect him from the alpha rays there. The suit is connected to the outside by a long plastic tunnel.

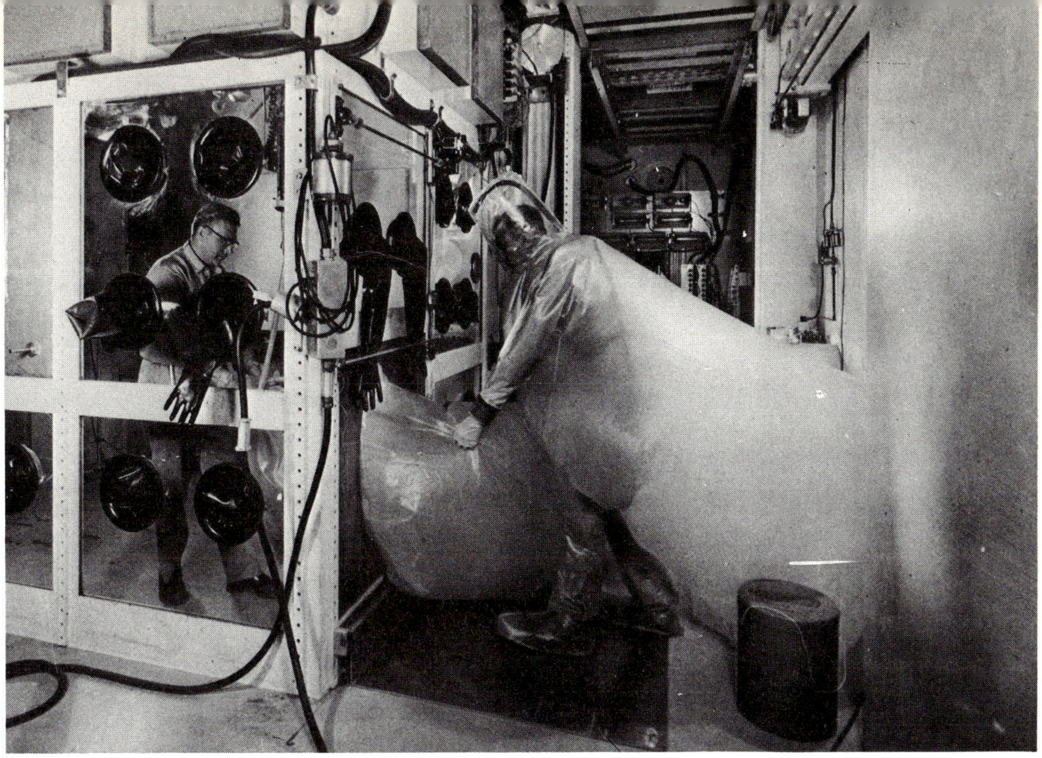

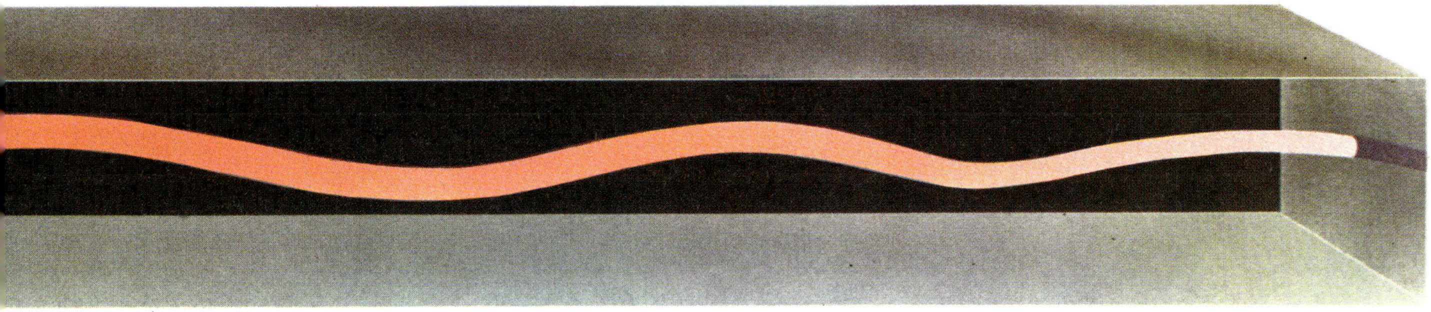

why you can only have a certain number of X-ray examinations every year and no more.

Protection from Radiation

Nuclear scientists and people who work at nuclear power stations all face danger from radiation. The materials they use are often radioactive. In fact, we all receive a small amount of radiation every day. This radiation comes from some minerals in the ground, and also from rays striking the Earth from outer space. But this natural radiation is very slight and is harmless.

Man-made radioactivity is far stronger than natural radioactivity. People who have to handle radioactive materials do so with special tools. They shield themselves behind thick screens that stop the radiation. Workers who have to go into radioactive areas wear special plastic

Below: A girl handling radioactive material wears a film badge on her lapel to detect any radiation in the laboratory.

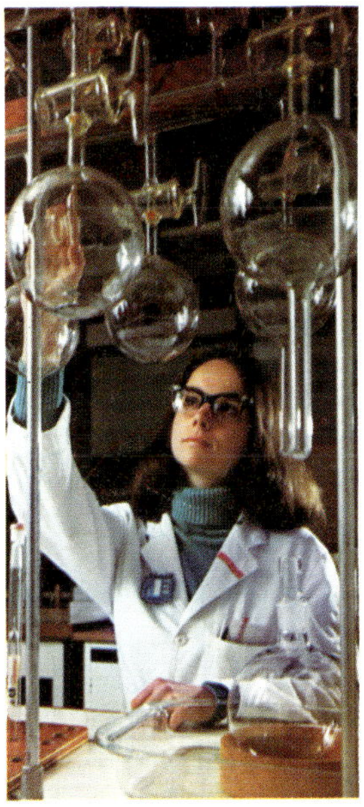

suits with their own air supply. The suits are washed down to remove radioactive particles before they are taken off.

All the people in nuclear power stations and nuclear laboratories have instruments to detect radiation. They each wear a film badge – a small piece of film inside a plastic holder. The film is developed and replaced with a new piece. If the wearer of the badge has received any radiation, the film appears black when developed. Other instruments that detect radiation include the Geiger counter. This instrument shows how strong the radiation is.

The designers of nuclear buildings try to make sure that no radioactive dust or other dangerous materials can leak into the air. Radioactive places, and containers of radioactive materials, are all marked with the radioactivity symbol to warn people to take care.

Putting Radiation to Use

We can make radiation work for us in all kinds of places – from the bottom of the sea to the Moon, from a factory to a living room.

Right: How radioisotopes are made. Isotopes give off alpha, beta or gamma rays – the three kinds of radiation. Radioisotopes are therefore a useful source of radiation, and making radioisotopes is an important industry. They are made by bombarding a piece of a non-radioactive isotope with atomic particles. In the picture, a piece is being placed in position by manipulators. A beam of particles is then produced by the cyclotron (right) to strike the sample and change it into a radioisotope. The operator cannot enter as the radiation is dangerous. The radioisotopes produced are packed in heavy boxes to prevent their radiation harming customers.

Left: Measuring the depth of snow automatically at a mountain pass in California. A radioisotope and a radiation detector move up and down the tubes. The changes in radiation received by the detector indicate the depth of the snow. The signals are sent along telephone lines to a central station. The information helps to prevent floods and to store water when the snow melts.

Right: The thickness of many products is controlled by passing radiation through the product to a detector. Here, cord for rubber tires is being examined.

Left: Light bulbs that need no electricity contain a weak radioisotope and a phosphor that glows when struck by the radiation. They never go out and are used to light signs and dials, such as this telephone.

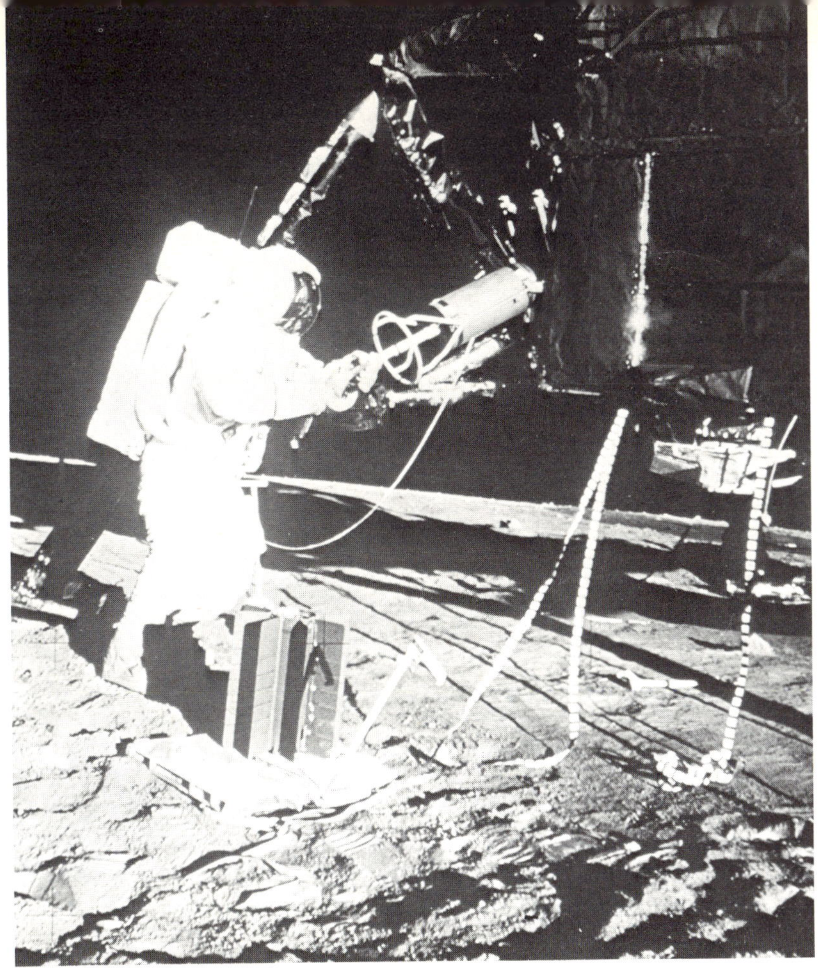

Above: Astronaut Alan Bean of Apollo 12 takes a canister of a radioisotope from his spacecraft. The isotope produces heat. When it is placed in the generator at his feet, electricity is produced to power the scientific experiments left on the Moon.

Above: Testing floor wax. Radioactive disks in a test floor are waxed. After walking over the floor, radiation levels show how well the wax wears.

Below: Examining an aircraft engine. Gamma rays from a radioisotope (left) pass through the engine intake and expose films around the outside (shown being placed in position). The pictures show up any faults in the engine.

Below: Radioisotopes are injected into mud that is to be dropped on the sea bed. Later radiation measurements show how the mud moves.

Atoms and Health

Radiation can be of great benefit to man – preventing, detecting and curing illness. It can also help us to learn more about how our bodies work.

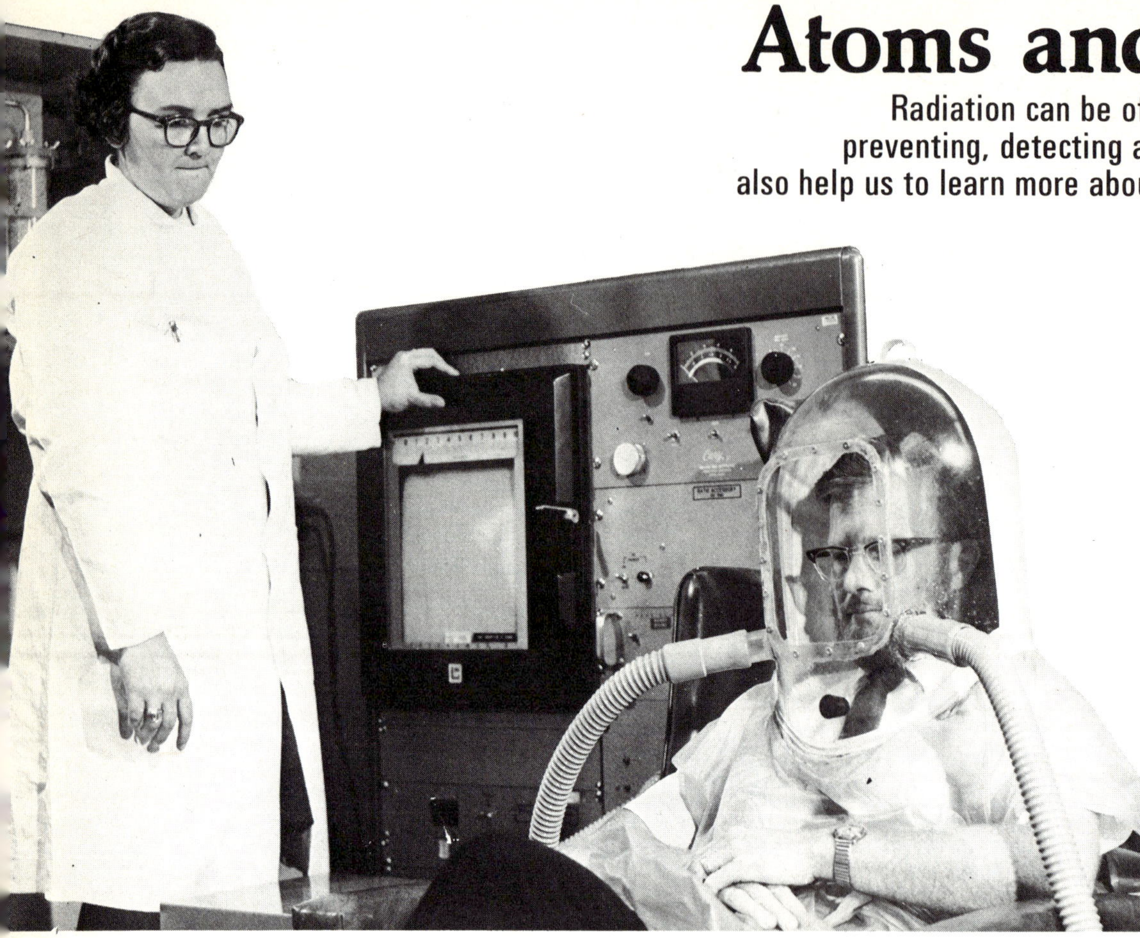

Studying The Body

Scientists inject small amounts of radioisotopes into living things. Then they trace their path through the body of the animal or plant by following the radiation it produces. In this way, they learn more about life.

Left: A scientist is given carbon-14 and his breath is then collected and analyzed. This tells us how the body uses carbon and helps in finding causes of illness.

Killing Germs

Right: Gamma rays are used to sterilize surgical instruments. Boxes of the instruments pass through a machine containing cobalt-60, which produces powerful gamma rays. Any germs on the instruments are killed by the rays, and the instruments are completely germ-free when they are unpacked.

Below: Potatoes 2 to 6 were given increasing doses of gamma rays, but potato 1 was not treated. The photograph was taken 35 weeks after the potatoes had been exposed to the rays. This shows that radiation sterilizes fruit and vegetables so that they can be stored for long periods.

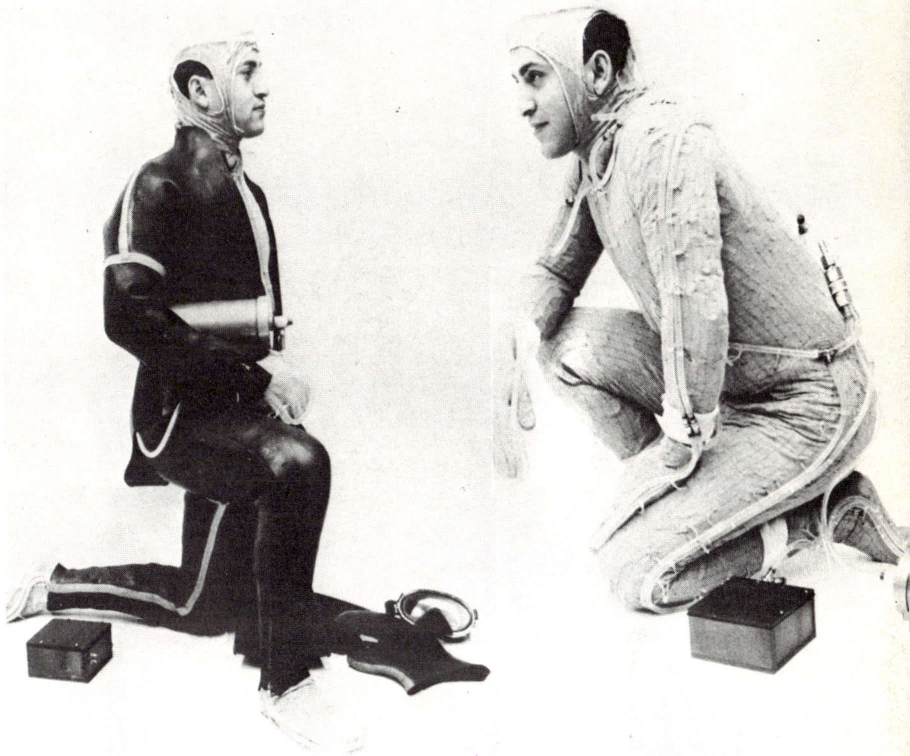

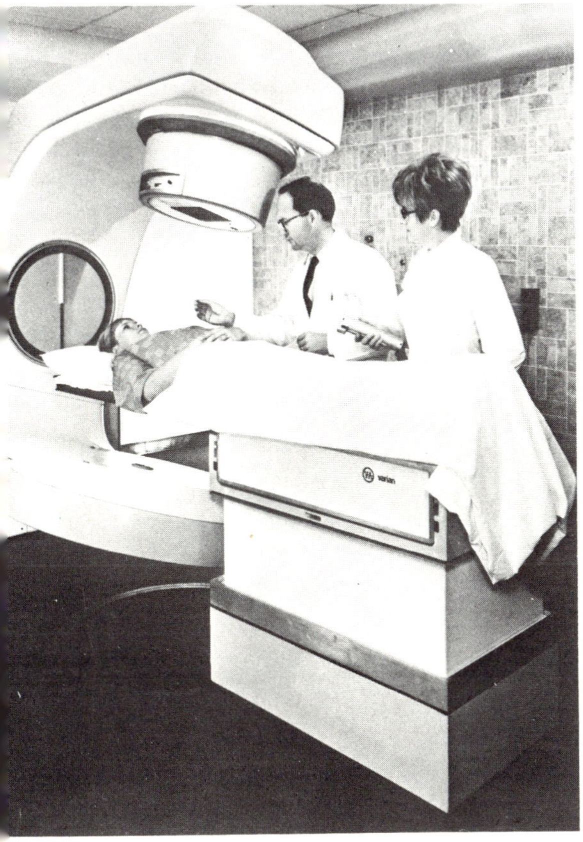

Giving Power

Above, left: A tiny battery powered by radioisotopes. It contains plutonium-238, which produces heat as it decays. The heat powers a generator. There is no radiation danger and the battery lasts for 10 years. It can power pacemakers, which keep people with heart trouble alive. The pacemaker produces electric signals that keep the heart pumping.

Above, right: A diving suit heated by nuclear power. A small heater containing plutonium-238 and a pump are attached to the suit. Warm water from the heater is pumped through tubes in a garment worn under the diving suit. By wearing heated suits, divers can work in cold water for long periods.

Curing Cancer

Below: Cancer tumors can be found by giving people radioisotopes and examining them with detectors. The tumor can then be cured by giving a strong dose of radiation (left).

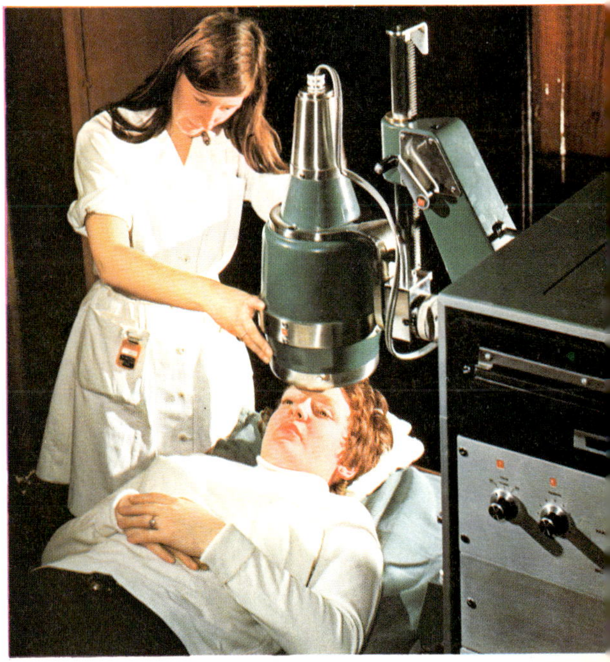

Into the Past

The speed at which atoms change can tell us how long ruins have stood and ancient men lived. It can even tell us the age of the Earth.

When atoms of a radioactive element shoot off alpha particles or beta particles, they change into different atoms. How quickly does this happen? Does the element suddenly change into another element, or does it happen slowly and bit by bit?

In fact, the speed of the change varies from one radioactive element to another. It is always measured by a quantity called the *half-life*. The half-life is the time it takes for half of *any* amount of an element to undergo change or radioactive decay. If the half-life of an element is one hour, for example, half of a sample of the element will change into another element in one hour. But the remaining half will not all change over the next hour. It must be thought of as a separate piece of the element, and half of it will change over the next hour – that is, a quarter of the

Above: Stonehenge, standing on a lonely plain in Wiltshire, England, is one of the world's finest prehistoric monuments. Radiocarbon dating of human remains show that the oldest part was built in about 2200 BC. But the age of a pick made from a deer antler, which was found in the ruins, shows that the high arches were erected about 400 years later.

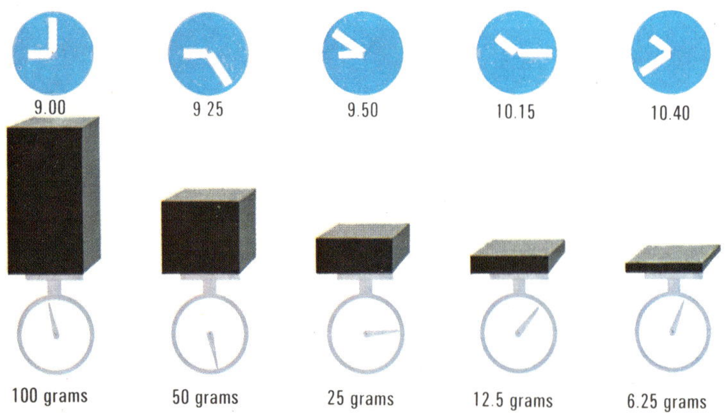

HALF LIFE

Suppose we had 100 grams of iodine-128, a black solid. This isotope of iodine is radioactive and changes by shooting off beta particles until it becomes xenon-128, which is a gas. The half life of iodine-128 is only 25 minutes, which means that half of the iodine would disappear every 25 minutes. If we put the 100 grams on a balance at 9 o'clock, only 50 grams would be left 25 minutes later, 25 grams another 25 minutes later, and so on. At 10:40, only 6.25 grams of iodine would remain.

original amount, leaving a quarter. Then, over the third hour, a half of this quarter will change, leaving an eighth, and so on. As every half-life period goes by, a sample of a radioactive element decreases by a half, then a quarter, an eighth, a sixteenth, a thirty-second, a sixty-fourth, and so on. There will always be a smaller and smaller amount left, until only one atom remains and this atom changes.

The half-lives of some radioactive isotopes are only a fraction of a second. But others, such as uranium-238 and thorium-232, last for thousands of millions of years. By measuring the amounts of these long-lived isotopes that are left in rocks, geologists can estimate how old the rocks are and even when the Earth was formed. The half-life of uranium-238 is 4500 million years, which is about the same age as the Earth. This means that there was twice as much uranium-238 in the Earth when it was formed. since then half of it has changed into other elements.

A radioactive element may change into another radioactive element, which then itself changes into another, and so on. Finally, one changes into an element which is not radioactive. Then the series stops. There are three natural radioactive series. They start with uranium-238, uranium-235 and thorium-232, and all end as lead after many millions of years.

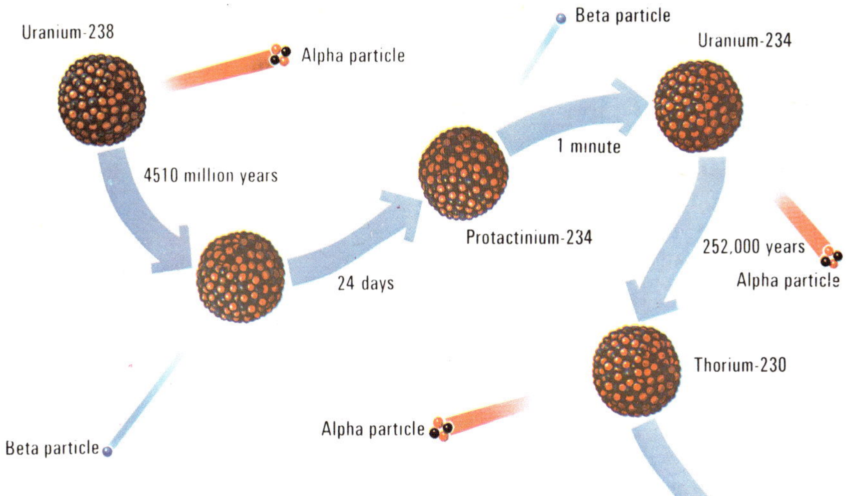

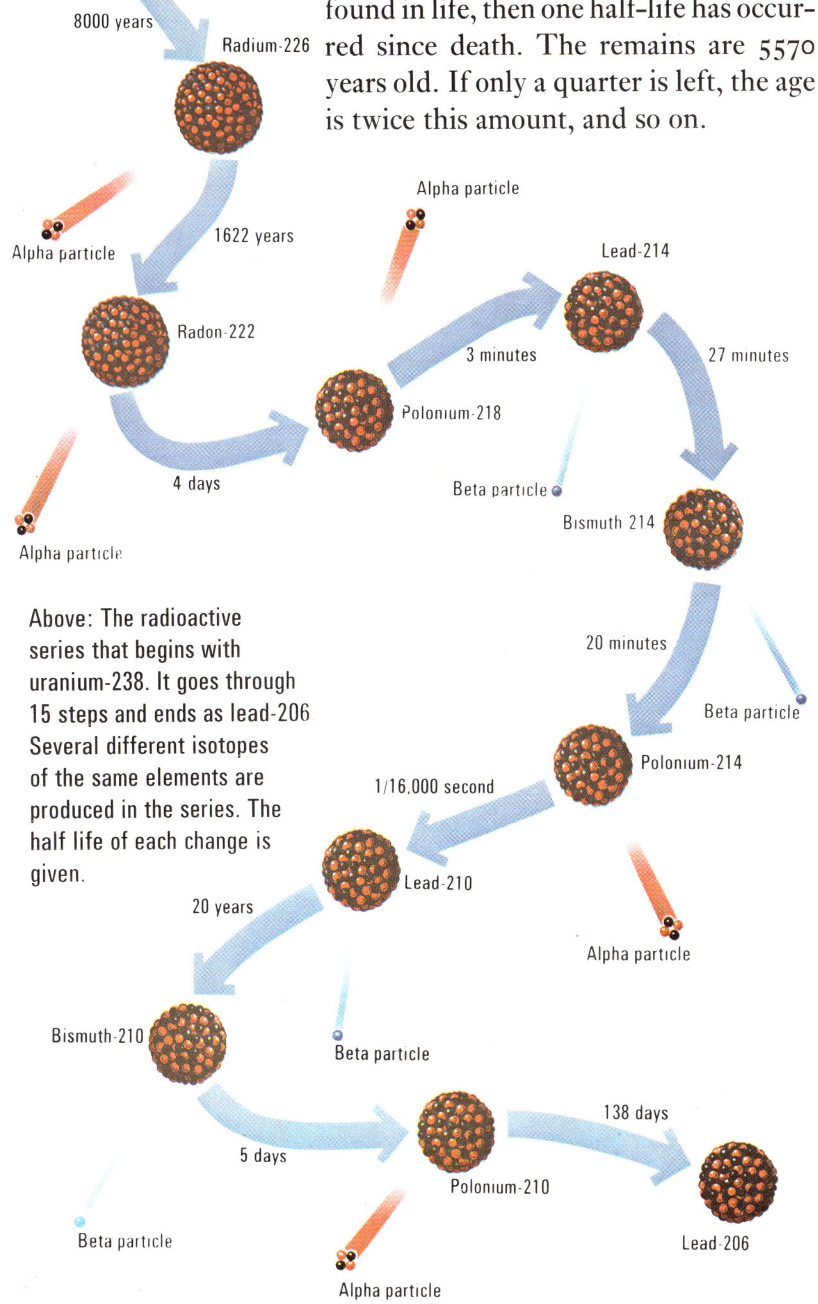

Above: The radioactive series that begins with uranium-238. It goes through 15 steps and ends as lead-206 Several different isotopes of the same elements are produced in the series. The half life of each change is given.

Radiocarbon Dating

People, animals and plants are all slightly radioactive, though not enough to cause any harm. They contain a fixed proportion of a radioactive form of carbon called carbon-14 or radiocarbon. When a person, animal or plant dies, no more radiocarbon is formed in the remains and the amount left slowly gives out beta particles and decays. Its half-life is 5570 years. The date when the person, animal or plant died can be found by measuring the amount of carbon-14 left in the remains. If there is only half the proportion found in life, then one half-life has occurred since death. The remains are 5570 years old. If only a quarter is left, the age is twice this amount, and so on.

Power from the Atom

When atoms break apart, they produce energy on a scale undreamed of by man.

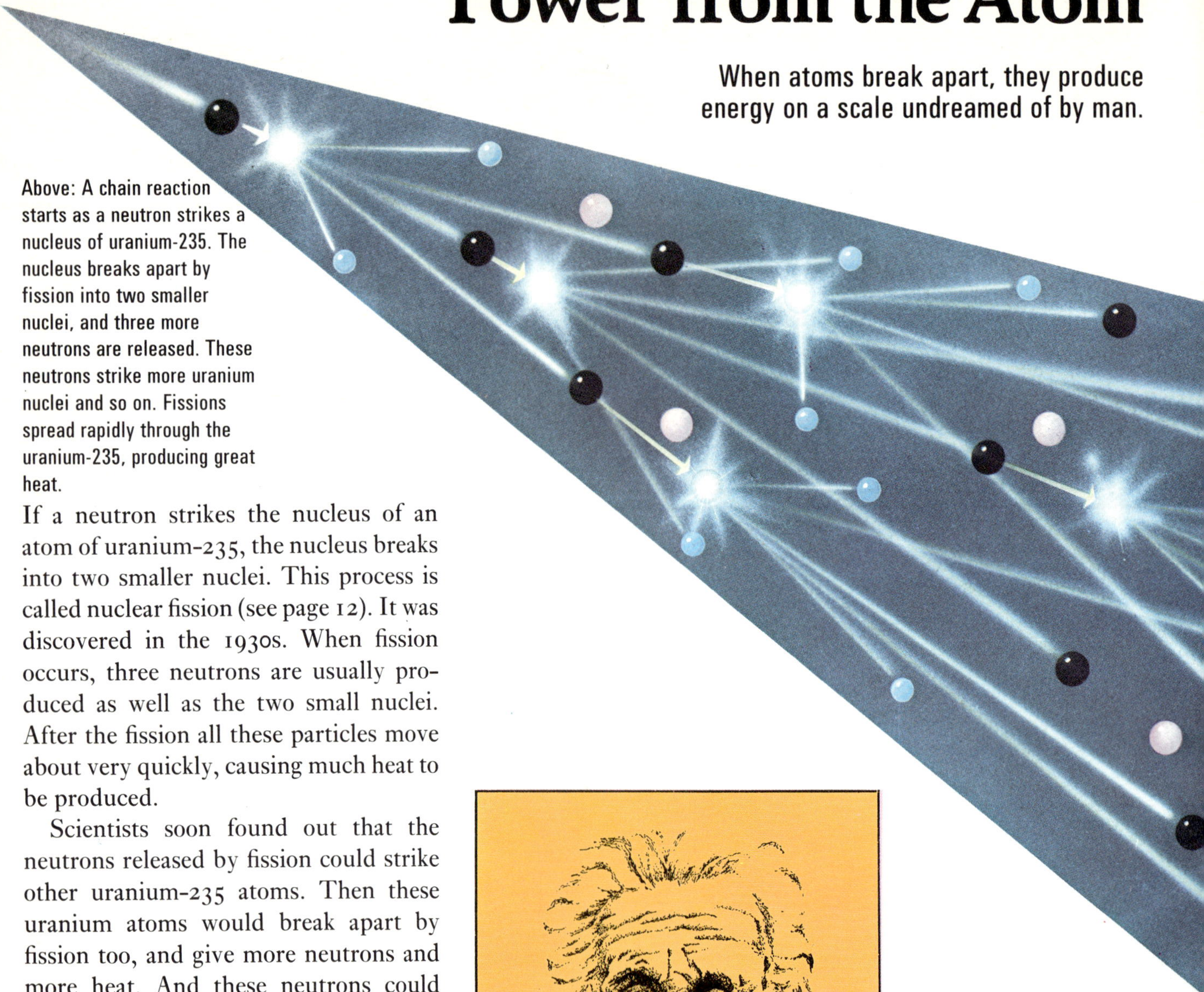

Above: A chain reaction starts as a neutron strikes a nucleus of uranium-235. The nucleus breaks apart by fission into two smaller nuclei, and three more neutrons are released. These neutrons strike more uranium nuclei and so on. Fissions spread rapidly through the uranium-235, producing great heat.

If a neutron strikes the nucleus of an atom of uranium-235, the nucleus breaks into two smaller nuclei. This process is called nuclear fission (see page 12). It was discovered in the 1930s. When fission occurs, three neutrons are usually produced as well as the two small nuclei. After the fission all these particles move about very quickly, causing much heat to be produced.

Scientists soon found out that the neutrons released by fission could strike other uranium-235 atoms. Then these uranium atoms would break apart by fission too, and give more neutrons and more heat. And these neutrons could strike yet more uranium-235 atoms and cause further fissions. With each fission the number of neutrons would increase. There would be a very rapid build-up in the number of fissions and in heat. This building-up is called a *chain reaction*. Scientists realized that if they could produce a chain reaction, they would have found a new way of making vast amounts of heat. The heat could then be turned into other forms of energy, such as electricity.

But a chain reaction does not occur in the uranium that is found in the ground. If it did, the world would have burned up long ago. Only one in every 140 atoms of

Albert Einstein

Albert Einstein (1879–1955) was a great German physicist. In 1905, he discovered that energy and mass are related by the equation $E=mc^2$. In this important equation E stands for energy, m stands for mass, and c^2 stands for the speed of light squared. This means that a little amount of any substance could be changed into a large amount of energy. Einstein's prediction came true when nuclear energy was first produced.

Uranium

Uranium is the main source of atomic energy. It has the greatest atomic weight of any element found in nature. The fission of a pound of uranium-235, a lump the size of a tennis ball, produces the same amount of heat as comes from burning nearly 1500 tons of coal.

uranium is a uranium-235 atom. It will undergo fission if struck by a neutron, but the neutrons produced are lost and do not reach other uranium-235 atoms. Therefore, a chain reaction cannot start in natural uranium.

To make a chain reaction, natural uranium must be treated to increase the proportion of uranium-235. A chain reaction can also be started in plutonium-239, which is made from uranium.

Above: This plant at Oak Ridge, Tennessee, produces uranium for nuclear power plants all over the world.

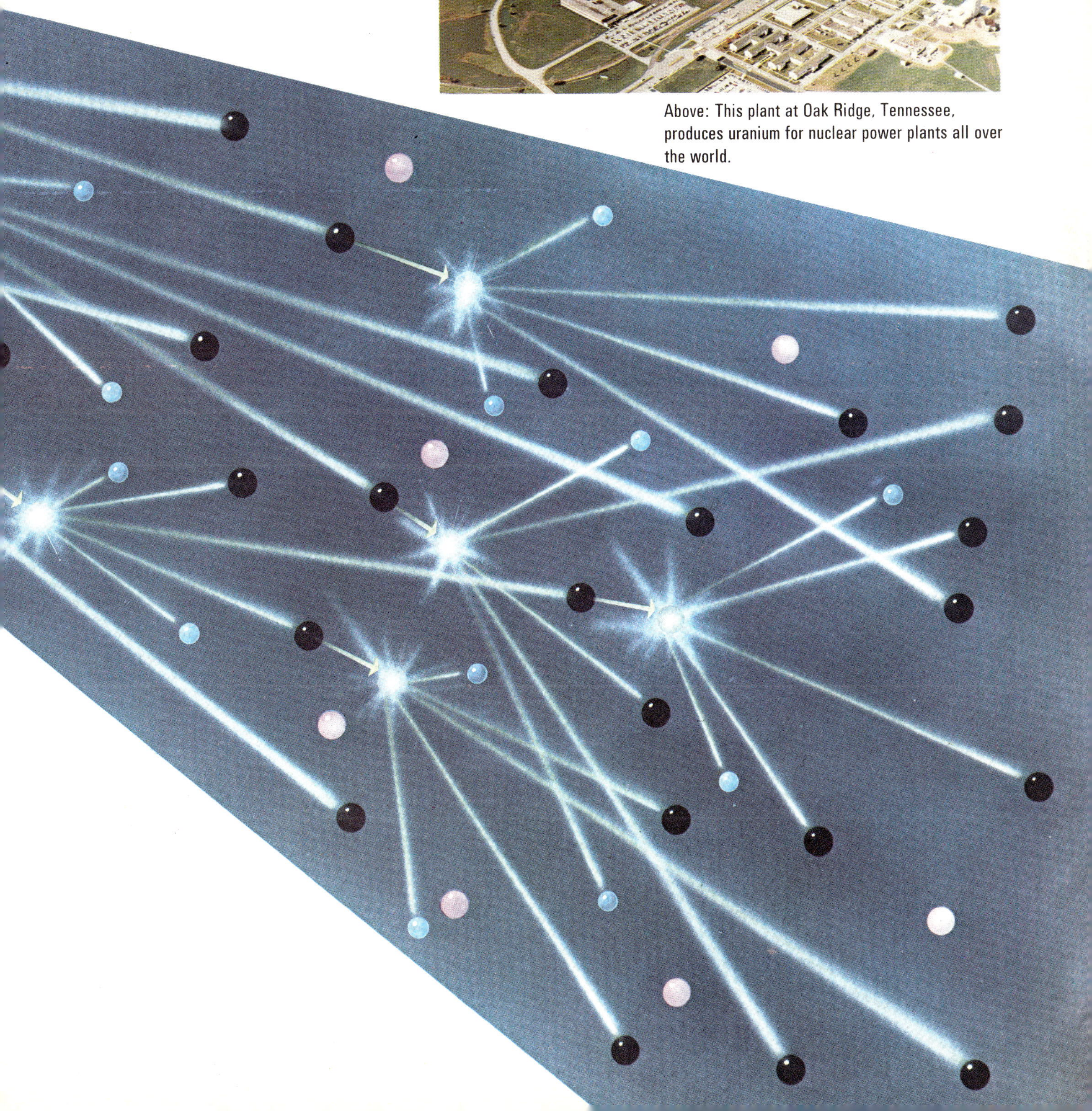

The Sun Touches the Earth

The atom bomb brought World War II to a swift end – but it also made the end of mankind possible.

July 16, 1945, is not a date that people think of as important. Yet on that day, something happened that was to be very important indeed. American scientists hid in shelters at a test site in the desert of New Mexico. They were waiting for a countdown to reach zero. Then a blinding flash of light lit up the desert. A mushroom-shaped cloud rose high into the air. Man had exploded his first nuclear weapon. It shook the test site with a thousand times more power than any bomb had ever done before. Its heat was so intense that it was as if the sun had touched the earth. The weapon was an atom bomb.

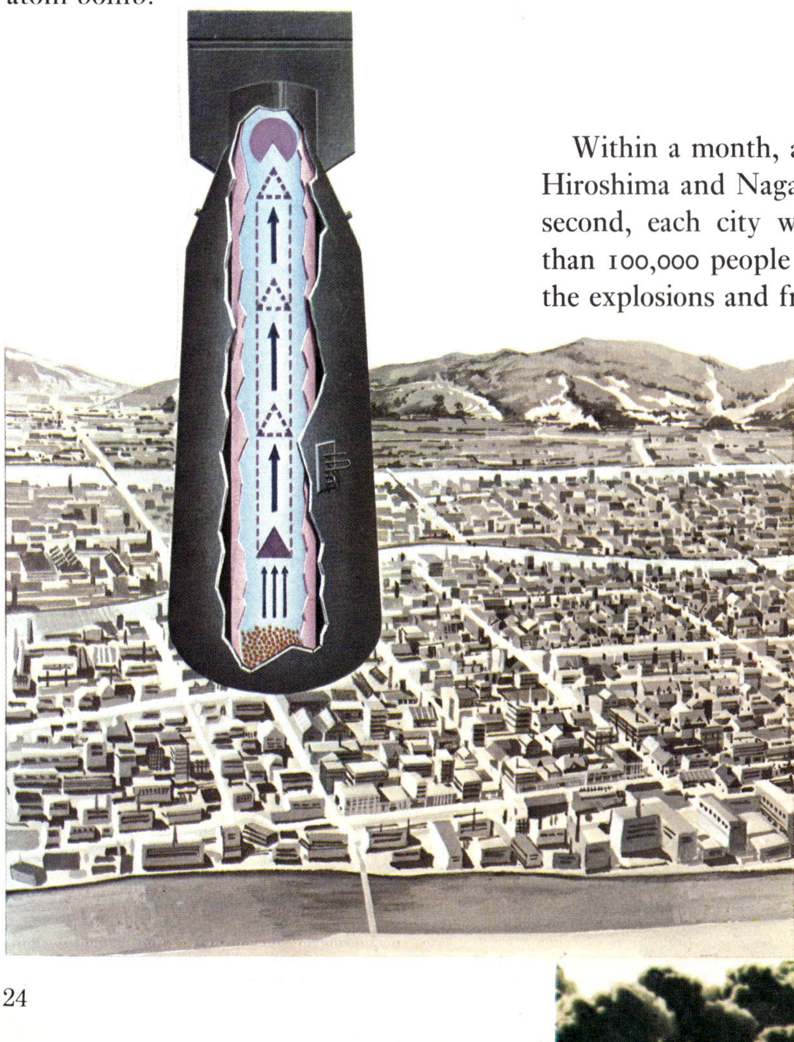

Within a month, atom bombs fell on Hiroshima and Nagasaki in Japan. In a second, each city was flattened. More than 100,000 people were killed – from the explosions and from the raging fires

Left: The atom bomb that fell on Hiroshima on August 6, 1945. The high explosive was detonated by a radar signal and drove two pieces of uranium together. A critical mass of uranium was produced, which immediately exploded with enormous force.

Right: An atom bomb test explosion. The huge mushroom cloud has become a symbol of the power of the atom used for destruction. Nuclear tests are now carried out underground by most nuclear nations.

that followed. But thousands more died later, victims of radiation sickness. The Japanese had no defense against such a weapon. They surrendered, thus ending World War II.

Atom bombs work by splitting nuclei of atoms in fission. They contain uranium-235 or plutonium-239. Scientists discovered during World War II that if a large enough amount of either of these substances were made, a chain reaction would start in it and it would explode.

The necessary amount of substance to produce a chain reaction is called the *critical mass*. To make a bomb, a few pounds are needed but, of course, it cannot be made in one solid piece, otherwise it would explode. Instead, several smaller pieces are brought together at the time of the explosion. In the Hiroshima bomb, two pieces of uranium-235 were forced together by high explosive (triggered by a radar signal) to produce a critical mass of uranium. The Nagasaki bomb contained a hollow ball of plutonium-239. To explode it, a high explosive crushed it into a solid mass and the chain reaction went off.

An atom bomb is one kind of nuclear weapon. The hydrogen bomb is another nuclear weapon that is so powerful that it needs an atom bomb just to trigger it off.

Below: A view of Hiroshima a month after the atom bomb exploded. The whole city has been destroyed, except for a handful of buildings.

The Nuclear Reactor

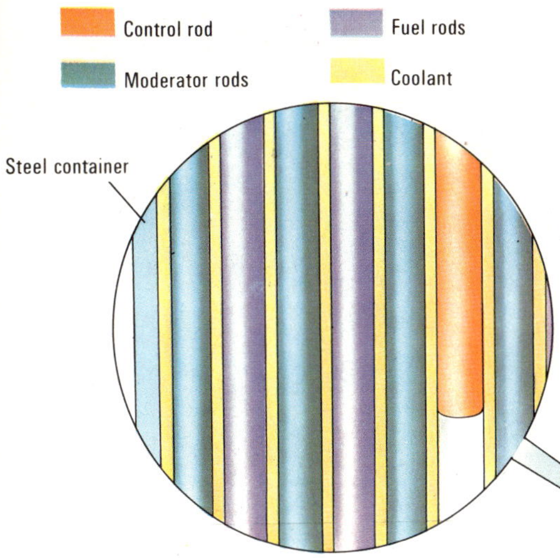

The Charging floor lies above the core of the reactor. Here, fuel rods are inserted into the core and removed after use. The control rods are operated from outside the core.

The Heat Exchanger takes heat from the core. Hot coolant flows through the exchanger and boils water flowing in tubes. This lowers the temperature of the coolant, which is then pumped back to the core. In some reactors, water is used as coolant and boils in the core.

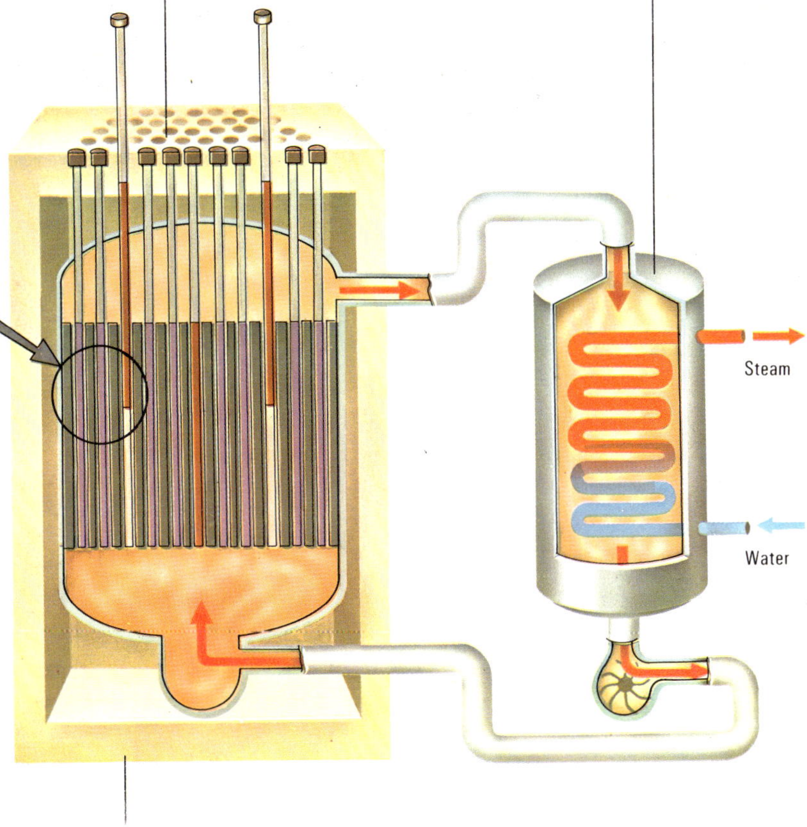

The Core of the reactor is built inside a strong steel container. It contains fuel rods made of uranium, inside tubes. The fuel rods produce heat. The control rods can be moved in or out of the reactor to vary the heat produced. The rods are surrounded by moderators. This slows neutrons produced by the fuel so that the chain reaction runs steadily. Through the core runs a coolant, a fluid or gas, to remove heat from the core. It is this hot coolant that supplies the energy to make electricity.

The Biological Shield is a thick layer of concrete round the core. It prevents the escape of harmful radiation.

some neutrons and slow the chain reaction. The reactor cannot explode, but it can get dangerously hot. In this case, the control rods can be pushed right into the core to stop the chain reaction. The core is surrounded by a thick layer of concrete to stop dangerous radiation and neutrons from escaping and harming people.

A cooling fluid or gas passes through tubes in the core to take up the heat. This hot coolant can then be used in various ways. Usually it goes to a heat exchanger, where it gives up its heat before returning to the core.

Making its own Fuel

When uranium-238 captures neutrons, it becomes plutonium-239, which can produce fission. This happens to a small extent in the fuel of reactors, but for plutonium to be produced in quantity, large amounts of neutrons are needed. This happens in a *fast reactor,* which contains fuel highly enriched in uranium-235 but no moderator. The reactor core is small but it produces large amounts of heat.

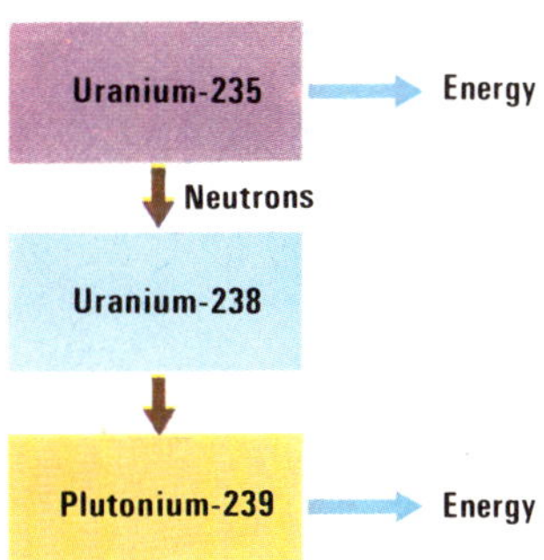

If a layer of uranium-238 is placed around the core, it is turned into plutonium-239. This plutonium can be used as a fuel for other fast reactors. This process is called *breeding* fuel. By breeding, 40 times more electricity can be made from uranium. Fast breeder reactors may be built in future instead of today's "burner" reactors. But many scientists believe that breeder reactors are less safe then burner reactors.

Nuclear Power Goes to Work

The silent nuclear reactor lies at the heart of many power plants. It can work day and night without refueling for months or years.

Above: The reactor building of a nuclear power station. Right: The power station from the air. Waste steam from the reactor building (left) goes to cooling towers (background) for recovery. The power generated is fed to an electricity substation (right).

USING NUCLEAR POWER

A reactor is at the center of every nuclear power system. It can produce plutonium or radioisotopes directly. Otherwise, the reactor's heat is used via a heat exchanger.

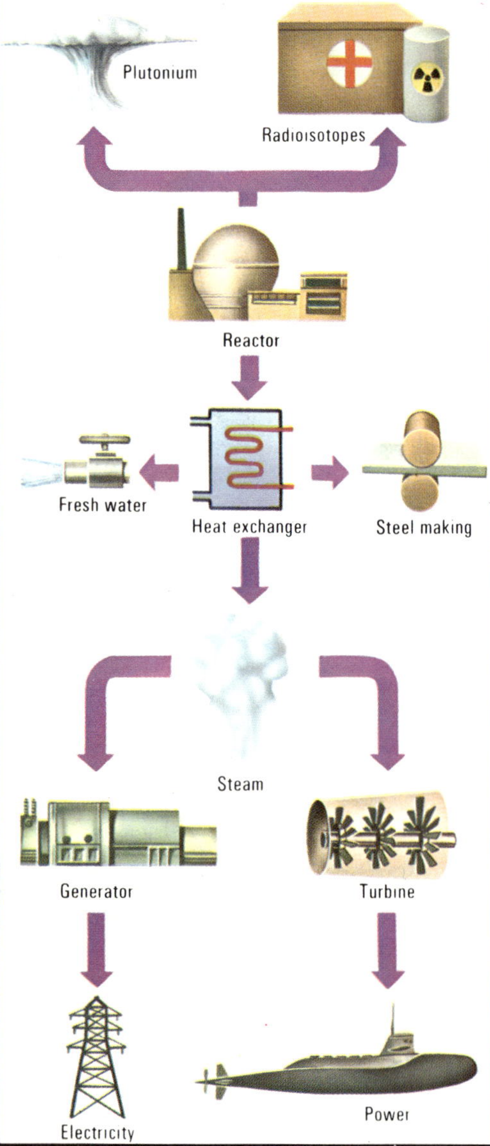

Fresh Water

Nations in many hot parts of the world have problems obtaining fresh water. Nuclear power is coming to their aid by providing better ways of producing fresh water from sea water. A heat exchanger heated by steam from a nuclear power station can be fed with sea water. The sea water evaporates, and the water vapor produced condenses to give fresh water.

Right: The world's first nuclear-powered icebreaker, ''Lenin''.

Below: A plant for making fresh water from sea water.

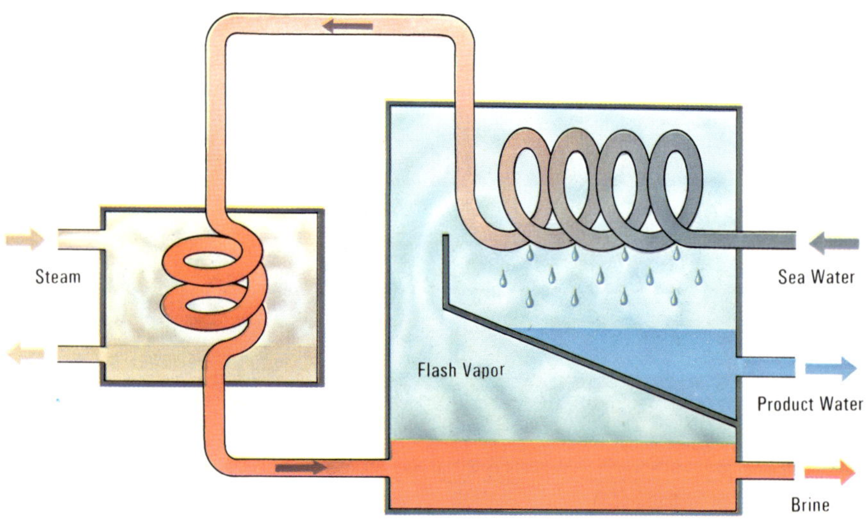

Electricity

Generating electricity is the main task of the nuclear reactor. Most developed nations have built nuclear power stations. By the year 2000 leading nations will probably be making about half their electricity by nuclear power. A nuclear power station has a reactor to produce heat in the form of hot steam from a heat exchanger. This steam drives the turbines of big electric generators, and the electricity produced is fed into the grid – the nationwide network of electric power cables.

When you are using electricity, you cannot know whether your power is generated by a nuclear station or an ordinary power station burning coal or oil. But the difference between the two kinds of power stations is obvious if you visit them. The nuclear station is clean, for nuclear power does not pollute the air. And it is silent, for the reactor requires only small amounts of fuel. It works for long periods before refueling is necessary.

Small nuclear power plants are vital in remote places, such as Antarctica. It is much easier to fly in small amounts of nuclear fuel to bases there than to transport huge quantities of oil.

Nuclear Ships

Nuclear submarines are vital to the navies of the most powerful nations. They are powered by nuclear motors. A small reactor produces hot steam, which is used to drive a turbine and power the propellers. The reactor does not need air, unlike the oil-fired engines of other submarines. So a nuclear submarine can stay underwater for months on end. Some surface ships are also powered by nuclear motors. They can travel several times around the world without refueling.

New Ideas

Reactor designers are looking at ways of using nuclear power in industry. One idea is that the heat from a nuclear reactor could be used to make steel in a blast furnace. Another idea is using the heat to make fuels from coal. Nuclear engineers are also working on ways of using hot gas from a reactor to drive a gas turbine. This would cut out the use of a heat exchanger. The power of the reactor would be used more efficiently.

Building Atoms

Energy is released when atoms come together to form larger atoms. It is this energy that makes the sun shine, and keeps us all alive.

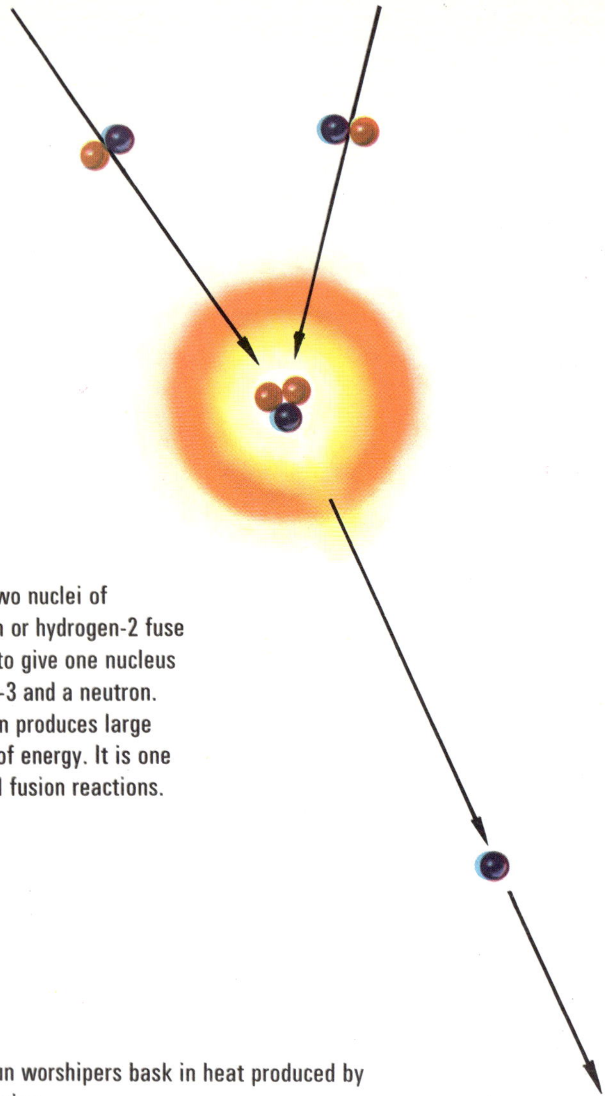

Above: Two nuclei of deuterium or hydrogen-2 fuse together to give one nucleus of helium-3 and a neutron. The fusion produces large amounts of energy. It is one of several fusion reactions.

When you lie in the sun on a sweltering summer day, you may wonder just why the sun is so hot. Indeed, the sun is so hot that the temperature at its surface (6,000°C) is high enough to vaporize even solid iron. But even this temperature is very small compared with the temperature at the center of the sun – about 13,000,000°C.

When any substance gets hotter, its atoms move about faster and faster. The sun is so hot that its atoms are moving very fast indeed. Often atoms collide with immense force. This force is enough to overcome the forces that usually keep atoms apart. The nuclei of atoms strike each other. Two nuclei may stick together so that their protons and neutrons form a new nucleus, and a proton or neutron is usually set free. The nuclei *fuse* together and what is called *fusion* occurs. *Light* elements such as hydrogen and helium are involved. (Fission is the splitting up of *heavy* elements such as plutonium or uranium.)

It is at first sight very strange that energy may be obtained both by splitting nuclei up and by joining them together. The answer lies in the fact that when any atomic nuclei get together there is always something which appears to be lost. This "loss" is the matter turned directly into energy.

Within the sun it is the turning of hydrogen into helium which makes the greatest amount of energy. Isotopes of hydrogen (one proton and three or more neutrons) combine together to

Below: Sun worshipers bask in heat produced by nuclear fusion.

form helium nuclei (normally two protons and two neutrons) so that vast stores of energy are released. This energy comes, of course, from the slight loss of mass of each proton and neutron involved.

The sun is a star, and astronomers have seen exploding stars in their telescopes. It seems likely that almost all the elements are made in these explosions. A few light elements, especially helium, are made as stars burn. Hydrogen itself is probably the starting material for the whole universe. All the other elements build up from it in the stars. So we depend on thermonuclear fusion for the very existence of all the atoms in our bodies.

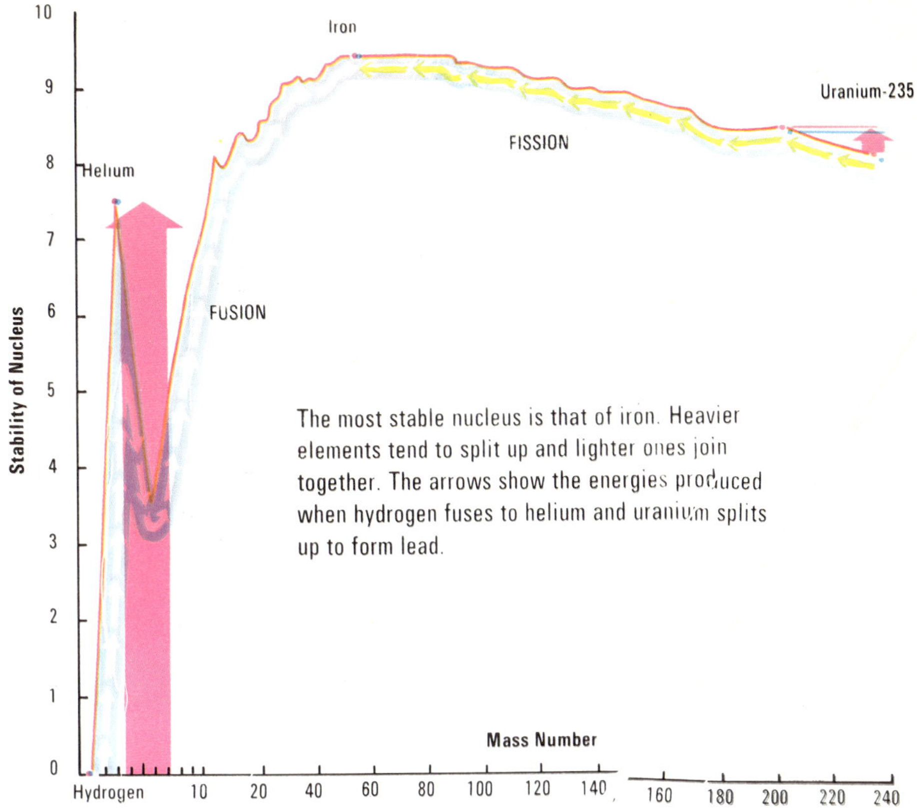

The most stable nucleus is that of iron. Heavier elements tend to split up and lighter ones join together. The arrows show the energies produced when hydrogen fuses to helium and uranium splits up to form lead.

Top right: Fission and fusion both release energy from the nucleus. Because the graph is steeper at the left we can see that fusion produces much more energy than fission.

Both fission and fusion can occur in a succession of steps. It is therefore possible that all elements may have been formed from hydrogen.

Right: The Crab Nebula in outer space is a cloud of hot gas and dust produced when a star exploded more than 900 years ago.

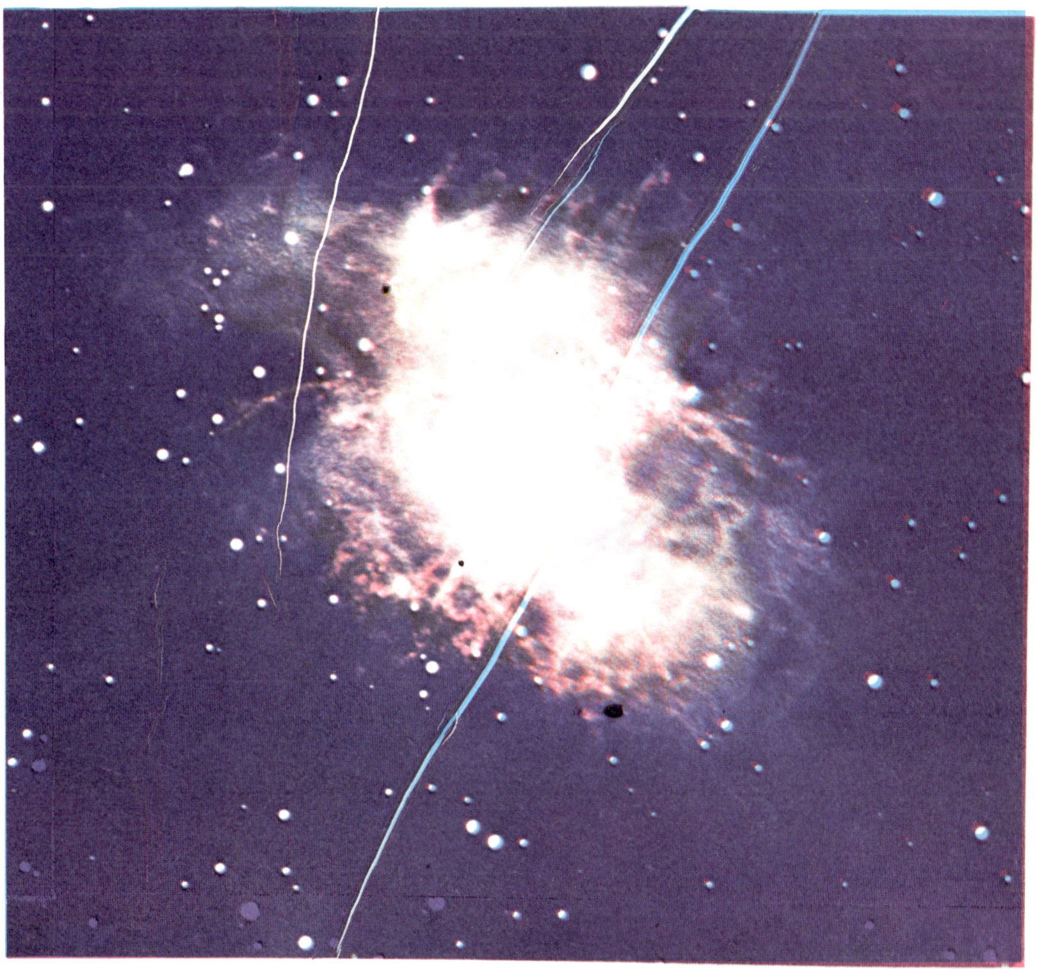

The Ultimate Weapon

Seven years after the first atom bomb, man developed a weapon so terrible that it uses an atom bomb as a trigger to set it off.

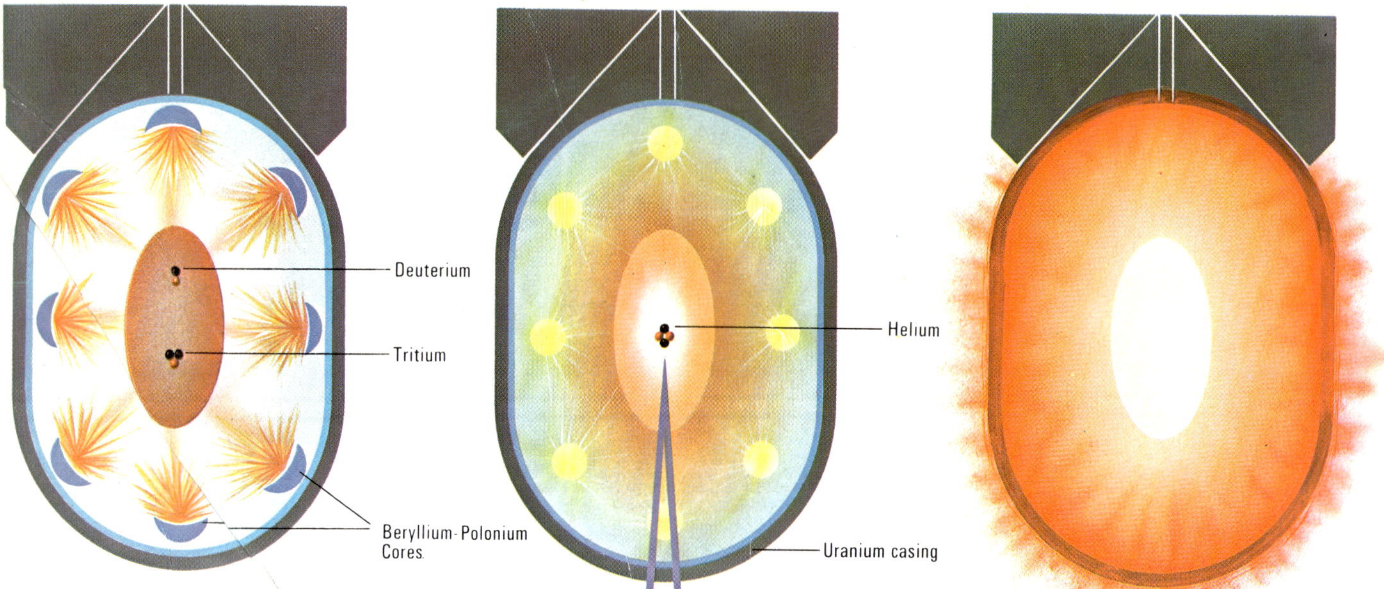

Above: How the hydrogen bomb works. Beryllium-Polonium cores trigger plutonium explosions which cause intense heat and pressure. Under these conditions, tritium and deuterium fuse to form helium. Neutrons from this fusion explosion set off the third stage – the fission of the uranium which forms the casing of the bomb itself.

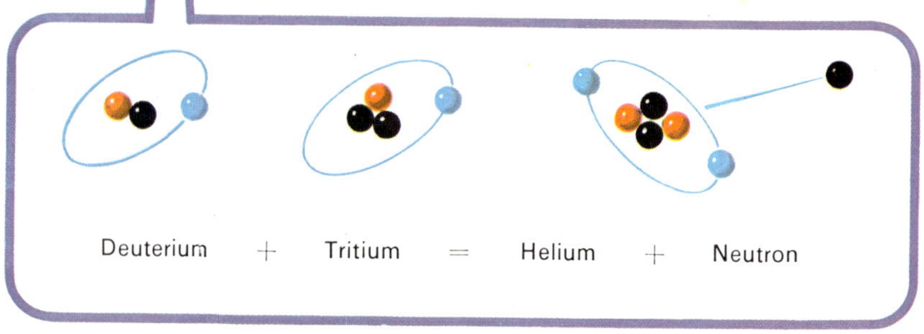

Left: A deadly mushroom cloud rises from a hydrogen bomb test. This photograph was taken 50 miles from the site two minutes after the bomb exploded. Ten minutes later, the cloud was 35 miles high and 100 miles wide.

Left: If a 10-megaton hydrogen bomb exploded over the center of New York, it would totally destroy the whole of Manhattan Island, which is 13 miles long. The rest of the city would be in flames, and people would be burned by the heat of the explosion as far as 30 miles away. Radiation sickness would cause death or suffering to many more.

There is a limit to the power of an atom bomb. It cannot produce an explosion greater than that of a million tons of TNT or one *megaton*. Soon after World War II, scientists began to wonder whether a bomb using fusion instead of fission could be made. Fusion works only at the temperature of the sun, and the only way to make such temperatures on earth is to explode an atom bomb. A fission explosion could be made to set off fusion in isotopes of hydrogen. Any amount of hydrogen could be used, and there is no limit to the power that could be produced.

The United States first tested a hydrogen bomb in 1952. It exploded with a force of 10 megatons – 700 times the power of the bomb that destroyed Hiroshima. The atom bomb trigger set off a fusion reaction in a mixture of deuterium and tritium, the heavy isotopes of hydrogen. Two years later, American scientists found an easy way to make the bomb even more powerful. They placed a jacket of uranium around the bomb. The fusion explosion was so powerful that it set off a fission explosion in the uranium. This second fission could raise the power at least five times.

Fission produces a deadly cloud of radioactive dust or fallout, but fusion does not. The uranium jacket around the hydrogen bomb made vast amounts of fallout. So these bombs came to be called "dirty" bombs. Scientists also worked to produce a hydrogen bomb without a uranium jacket. This bomb can produce almost all its power by fusion and produces very little fallout. It has been called a "clean" bomb, but no hydrogen bomb is really clean.

In addition to the United States and Russia, Britain, China and France have tested hydrogen bombs. Although hydrogen bombs have never been used in war, the bombs that these nations possess are a deadly threat to mankind.

Power Unlimited

If we can harness the power that makes the sun shine, then man's power problems will be solved.

Soon after the first atom bomb exploded scientists were able to build nuclear power stations. But it is now more than 20 years since the first hydrogen bomb shook the earth, and we have not been able to use fusion for anything other than weapons. However, scientists in many countries are hard at work trying to harness fusion power. Their aim is to build a reactor in which deuterium nuclei can be made to fuse together. Deuterium, or hydrogen-2, is a heavy isotope of hydrogen. One atom in every 6000 hydrogen atoms is a deuterium atom. Deuterium can be obtained easily by passing an electric current through water. And water is so abundant on earth that our supply of deuterium is almost unlimited. A pound of deuterium could produce as much energy as 1000 tons of coal. If we can control fusion, then fuel shortages and power problems will be things of the past. But to produce fusion, the nuclei of atoms must come together. It needs a very high temperature to give them enough energy to meet. There must also be a certain number of atoms in any one space; that is, the pressure

Above: The sea may one day provide us with a supply of fuel that is almost unlimited. For every 6000 hydrogen atoms in the water, there is one atom of deuterium, an isotope of hydrogen. If the immense power of the hydrogen bomb and the sun can be controlled in fusion reactors, deuterium will be the fuel.

must not be too low. Also, fusion is more likely the longer these two conditions are kept up.

These are the problems. The temperature of the deuterium is measured in millions of degrees. It is so high that no ordinary container can hold it. The container walls would cool the deuterium too much. And it is also difficult to keep the pressure up for long enough. Scientists have built vessels in which the deuterium is held in a magnetic field, so that it does not touch the walls. The magnetic field also squeezes and heats the deuterium. But the deuterium moves out of the magnetic field before it is hot or compressed enough to produce fusion. If fusion happens, the reactor will make more energy than is put into it to make the fusion happen. This energy will appear as heat, as in fission reactors, and also directly as electricity.

In the early 1970s, military scientists began to try out new ways of producing fusion in order to make small nuclear weapons. This new idea is based on the laser. (A laser is a beam of light so intense it can cut through metal.) By firing laser beams at small pellets of solid deuterium, fusion of the pellets might occur. The laser beams would heat the outside of the pellet instantly to a high temperature. This rapid heating would place the center of the pellet under intense pressure. A small fusion explosion could possibly occur – not enough to wreck a reactor but enough to produce immense amounts of heat.

Scientists are pressing forward in all these ways to try and make a fusion reactor. The research is becoming so expensive that nations are beginning to cooperate with each other to achieve what is likely to be one of man's most important discoveries.

Below: Scientists working on a machine for a fusion experiment. In the circular tube, a beam of deuterium is squeezed by a strong magnetic field. So far, no one has been successful in harnessing fusion power. But scientists in many countries are hard at work trying to solve this vital puzzle.

Below: Controlled fusion may be achieved by firing laser beams at pellets of nuclear fuel.

Nuclear Energy: Boon or Threat?

The immense power of the atom promises great benefits to man, but unless we treat it wisely it can do us more harm than good.

Supplies of coal and oil are not unlimited and our demand for energy is always increasing. Many nations have to import large amounts of fuel such as oil to keep up industrial production. These countries have looked for other ways of making enough energy. Producing electricity by nuclear fission is the only way that seems possible at the moment. So nuclear power programs are expanding rapidly. Without power generated by fission, the lives of many people would not be so comfortable. An extra benefit is that nuclear power stations do not pollute the air as coal- or oil-fired stations do.

Nuclear power helps us in other ways. The production of radioisotopes has helped industry and medicine, and new uses are being found for them all the time. And perhaps nuclear explosions could have important peaceful uses. Underground explosions could help break up minerals and make them easier to mine. Canals could be dug and harbors blasted in solid rock more easily with nuclear explosives than with ordinary explosives. The explosives can be made to produce little radioactivity. But more research is needed to make them safe.

Safety is very important when we use nuclear energy. The use of nuclear energy always releases radiation, and radiation is very harmful to life. Unless nuclear energy is strictly controlled, it can be a grave threat to us all. One problem is that it is difficult to find out how much radiation people can absorb without harm. The longer we live with nuclear energy, the more we find out about the effects of radiation. And it seems that quite small amounts of radiation may have long-term harmful effects. Some scientists believe that increases in child deaths may be caused by the little increases in radiation that have occurred since man began to use nuclear energy. Although safe limits are published for radioactivity, it is possible tht no amount of radiation, however small, is harmless.

Below: Nuclear explosions a mile or more underground produce large cavities of broken rock from which minerals, oil or gas might easily be obtained.

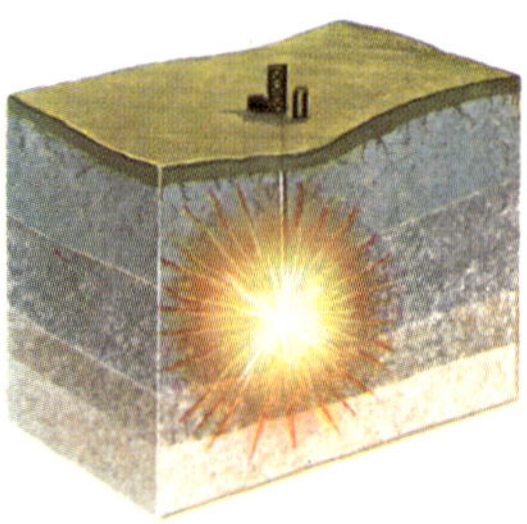

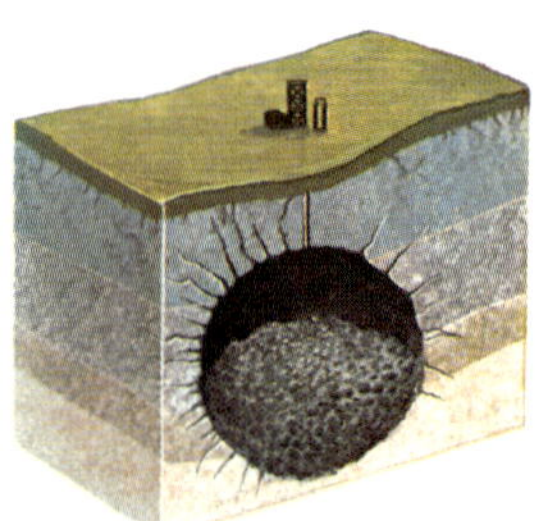

Radiation may enter our bodies from many sources. A little leaks from nuclear power stations, however strong their safety measures. There have been small accidents at power stations that resulted in some radioactivity being released. An ordinary reactor cannot produce a nuclear explosion. But if the reactor should get too hot and break up, radioactivity would be released for miles around. The new, fast, breeder power stations look more dangerous. These stations operate at high temperatures. Experts believe that safety measures are not sufficiently good to make an accident impossible.

Controlling the Atom

In a nuclear reactor the atom bomb is tamed to produce energy we can use. Some reactors can even make their own fuel.

An atom bomb explodes because the chain reaction gets out of hand. Each fission of an atom produces neutrons that cause two or three more fissions. These produce more neutrons that cause even more fissions and so on, all in the fraction of a second. But the chain reaction can be slowed so that only one neutron from a fission causes another fission. Then the reaction proceeds steadily and does not multiply. This happens in a nuclear reactor. It is made so that a runaway chain reaction and a nuclear explosion are impossible.

In natural uranium, only one in every 140 atoms is uranium-235. The other atoms are uranium-238, which does not produce fission when struck by neutrons. A chain reaction does not occur in natural uranium because there are not enough uranium-235 atoms to keep it going. Also, fission is more likely to happen if the neutrons move more slowly. A nuclear reactor can therefore work if the natural uranium is treated to increase the amount of uranium-235, and if the neutrons produced are slowed down.

The fuel for reactors therefore usually consists of *enriched* uranium or uranium compounds. These contain a higher proportion of uranium-235. The fuel is placed in long tubes, which fit into the *core* – the center of the reactor. To help make the chain reaction go, a *moderator* is placed around the fuel to slow the neutrons. The moderator is usually graphite or water. In case the chain reaction starts to go too quickly, *control rods* are also inserted into the core. They are made of metals that absorb neutrons, and can be moved in or out of the core.

Above: A view inside the core of a nuclear reactor. The tubes contain fuel rods, control rods, and a moderator.

As the chain reaction goes on, heat is produced. If the core gets too hot, the control rods are pushed in to take up

Enrico Fermi

Enrico Fermi (1901–1954) was an Italian physicist who built the first nuclear reactor. In 1934, Fermi became the first man to produce fission. He left Italy for the United States and began to build a nuclear reactor at the University of Chicago. The reactor came into operation on December 2, 1942, and man entered the nuclear age.

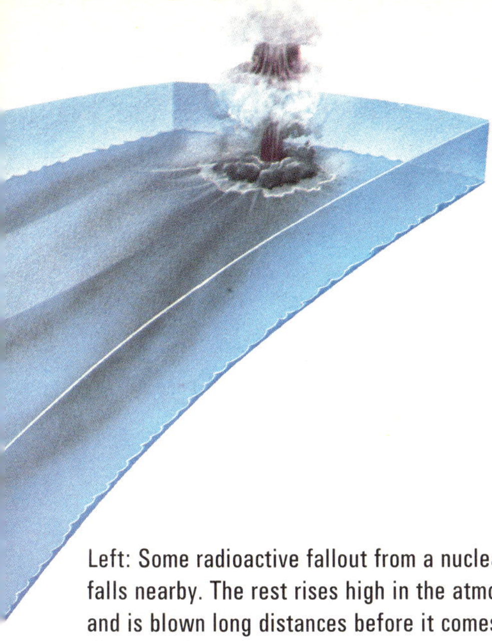

Left: Some radioactive fallout from a nuclear test falls nearby. The rest rises high in the atmosphere and is blown long distances before it comes down to the ground.

Below: A man is dwarfed by huge bins in which nuclear waste is stored. Getting rid of increasing amounts of dangerous waste can be a serious problem.

The waste that nuclear stations produce is very radioactive. Every *day*, a large power station makes as much radioactivity in waste as four of the bombs that dropped on Hiroshima. By the year 2000 the waste problem will be much worse. The amount of waste produced every day will be about the same as 2000 Hiroshima bombs. The volume of waste is not large – a big station produces only about half a gallon – but its radioactivity is intense and very dangerous. Nuclear waste is usually stored in tanks with thick walls that absorb the radioactivity. But tanks corrode and may after a while split open. And radioactivity can last for thousands of years. Nuclear authorities are finding out about dumping as a way of getting rid of nuclear waste. It might be buried in the ground or dumped into the sea. But the movements of ground and sea are too uncertain to make this kind of waste disposal completely safe. As nuclear power increases, waste disposal will become a vast problem. Already, the more developed countries have their own government bodies to control the dumping of radioactive wastes.

Dangerous Fallout

Fallout from tests of nuclear weapons has already threatened our health. Radioisotopes from bomb tests are blown around the world high in the atmosphere. After they descend to earth in rain, they find their way into our food, water and milk. Then they concentrate in the body. Fallout, even thousands of miles from a test site, exposes us all to a little extra radiation. As we know so little about the effects of radiation, it could do us harm. Illnesses such as leukemia have been linked with fallout.

From 1945 to 1963, the United States, Russia, Britain and France exploded nuclear bombs in tests. Five hundred tests were held, and the fallout from them began to worry governments. In 1963, a treaty signed by most of the world's nations banned tests in the atmosphere and on the ground. Most nuclear tests have since been held underground. These do not put radioactivity in the air unless there is a leak, which is very unlikely. However, France and China took no notice of the ban and continued to poison the air with fallout from surface tests. A further danger is that other nations will follow the example of France and China and start nuclear tests.

One great danger that all governments fear is that terrorists might get hold of uranium-235 or plutonium-239 and make an atom bomb. They could use it to hold a government to ransom, and it could explode by accident or even perhaps on purpose. Measures are taken at nuclear laboratories and stations to stop theft. But these measures may not work if a group is very determined to steal nuclear fuel. One way of avoiding this danger would be to make only a kind of plutonium that can be used in reactors but not in weapons. This could be done by making changes in the design of fast reactors.

Organizing the Atom

By sharing their knowledge of nuclear energy, nations can advance its peaceful uses.

Above: An IAEA scientist examines a new strain of barley made by exposing crops to radiation.

Left: In an IAEA laboratory, rice roots are fed with radioisotopes to help find how rice grows.

Right: Pests can be controlled by sterilizing insects with radiation. The sterile insects fail to breed. Here, an IAEA biologist examines apricots for damage by pests.

The International Atomic Energy Agency (IAEA)

The IAEA was set up in 1956 by the United Nations to help promote the peaceful uses of nuclear energy. Its headquarters are in Vienna, Austria. More than a hundred nations are members, including all the nations that have the most knowledge of nuclear energy. The IAEA is vital to poorer nations who wish to find out more about nuclear energy and to make use of its benefits. The agency provides scientific knowledge through its information service and books. It organizes conferences, gives advice and assistance where needed, and it sets up research projects. For example, IAEA scientists are studying the movement of underground water in dry areas by tracing natural isotopes. Their results will improve water supplies in these regions.

Below: The symbol of the IAEA.

The IAEA carries out research into all the peaceful uses of nuclear energy. These researches cover food and agriculture, medicine and radiation protection, and nuclear waste disposal. It has set up standards to be followed wherever nuclear energy is used.

One very important role, undertaken by the IAEA and other nuclear energy organizations is to make sure that nuclear fuel from power stations and laboratories is not used to make weapons or explosive devices. IAEA inspectors check that all the fuel is used for peaceful purposes.

European Organizations

Euratom is the nuclear organization of the nine countries of the European Economic Community (EEC). Its headquarters are in Brussels, Belgium. Euratom carries out research into nuclear reactors. It is also interested in the medical uses of nuclear energy, and other peaceful uses. It publishes information and gives advice on nuclear matters. Euratom also arranges for the supply of nuclear fuel to its members.

CERN is the name of the European Organization for Nuclear Research. It has its headquarters at Geneva, Switzerland. CERN carries out research into nuclear physics, a costly business whose finance is shared by 12 European nations. CERN's work is concerned with the particles that make up the atom rather than the use of nuclear energy. But CERN's discoveries may well be of use in nuclear energy research.

Right: A view of the headquarters of CERN, the European Organization for Nuclear Research, near Geneva. The dotted circle shows the site of a huge new underground particle accelerator. The border between France and Switzerland is marked by small crosses.
Below: A physicist in a European laboratory in Italy studies a method for turning heat directly into electricity. This could make nuclear power stations produce much more electricity.

The *Nuclear Energy Agency* is an agency of the Organization for Economic Cooperation and Development (OECD). Its headquarters are in Paris. Most European nations are members, as well as Japan, and the United States and Canada. Its aims are similar to those of the other nuclear energy organizations.

Atoms in Space

Man's future exploration of other planets in the solar system will probably depend on nuclear-powered engines.

Interplanetary space is not completely empty. Atoms and particles of atoms are scattered thinly between worlds. At the center of the solar system the sun is always producing a stream of protons and electrons as hydrogen atoms break up. They move around the Earth, producing belts of radiation beyond the atmosphere. Astronauts on their way to or from the Moon move quickly through the radiation belts and avoid harm. If the particles come any nearer Earth, they enter the atmosphere. Then they strike atoms in the air and produce the bright glow known as the aurora, or northern and southern lights.

Right: A design for a manned spaceship to reach Mars. The engines are nuclear-powered.

Below: The aurora that lights the night sky is made by electrified atoms striking our atmosphere.

Nuclear Power in Space

The first nuclear reactor to leave the Earth was placed in orbit in 1965 aboard a U.S. satellite. It produced electricity to power the systems of the satellite. Nuclear generators produce a lot of power, but they are heavy. For satellites, it is usually better to rely on solar cells, which turn sunlight into electricity. But one day, nuclear-powered satellites may be broadcasting television programs directly into our homes from space.

Rockets of the Future

Today's space rockets use chemical engines that burn fuels. The hot gases produced fly from the engine's exhaust. As they do so, they push against the rest of the engine (just as you push against the ground to jump into the air). This push sends the rocket forward. To give the rocket more power, the gases must leave the exhaust at higher speeds.

Above: The Jupiter spaceship from the famous film *2001: A Space Odyssey*. At the rear of the ship is the nuclear engine. It is placed far from the crew compartment to lessen dangers from radiation.

A nuclear rocket engine can be designed to give higher exhaust speeds. Gas is heated by a nuclear reactor to very high temperatures. The atoms in the gas move very fast at these temperatures, and so give a nuclear rocket great power. One advantage of a nuclear rocket is that no particular kind of gas is needed. Any kind could probably be used, and the rocket may even be able to refuel on a distant planet. At present, nuclear engines are too large and heavy to be used in space travel.

Another kind of future rocket engine is the electric engine. In this engine, atoms of gas are charged electrically and then speeded up to about 50 miles a second by applying an electric field to them. Electrified atoms are called ions, and so this kind of engine is known also as an ion engine.

Ion engines produce very little push or thrust, but they can operate for long periods. Starting from Earth orbit, a rocket powered by an ion engine would slowly accelerate over a long period to an immense speed. The electrical power for the engine would come from solar cells, or, if the rocket were far from the sun, from a nuclear generator.

An Ion Engine contains a source of electricity and a propellant gas. Atoms in the gas are first electrically charged to form a beam of ions. The beam is then speeded up by an electric field. As the beam leaves the engine at high speed, its electrical charge is made neutral by adding electrons to the beam.

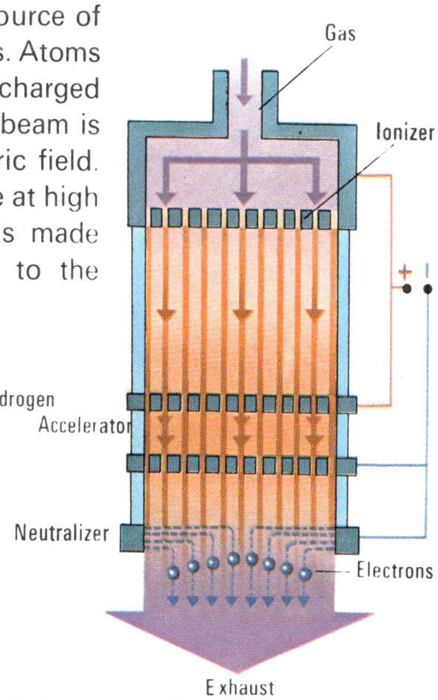

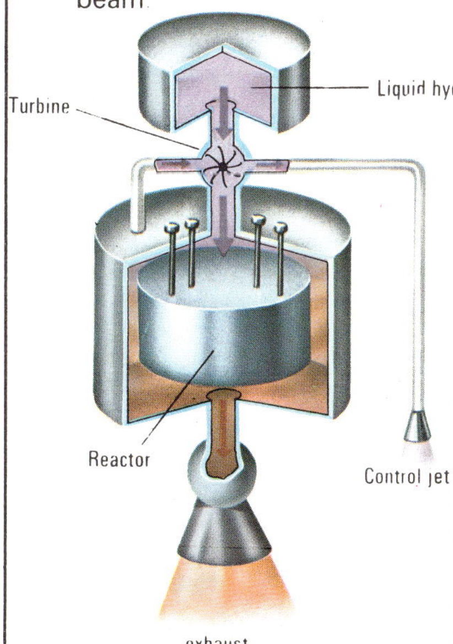

A Nuclear Engine contains a reactor that heats liquid hydrogen fuel. Most of the hot hydrogen gas leaves the engine exhaust at high speed. The rest of the gas is used to drive a turbine to pump fuel from its tank.

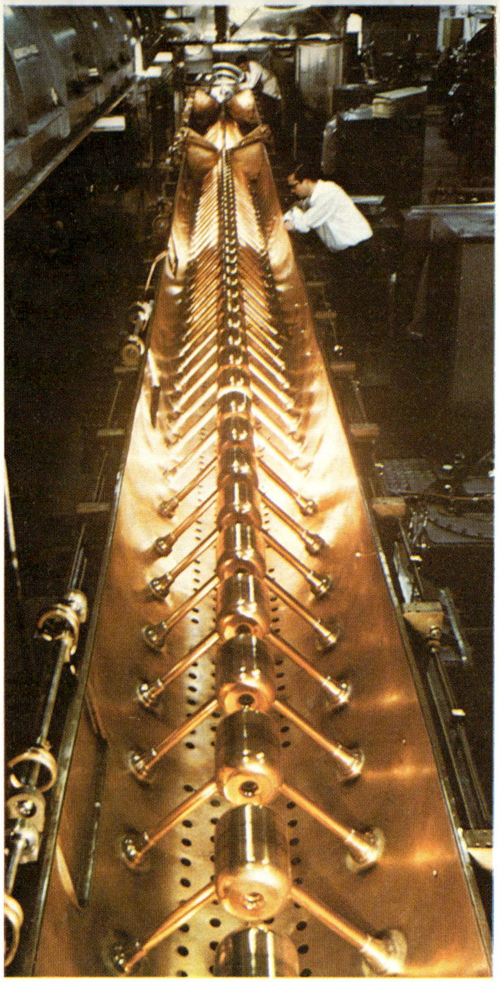

Above: The drift tubes are exposed to view when a linear accelerator at CERN is opened.

PARTICLE ACCELERATORS

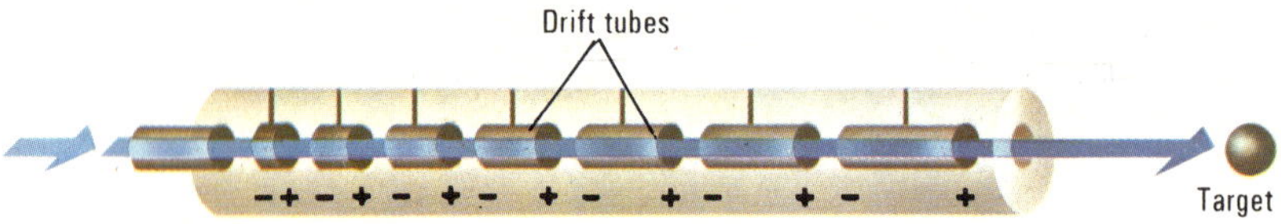

In a Linear Accelerator, particles are sped by electric charges at the drift tubes. Each tube is charged positive and then negative so that the particles are pulled towards the tube and then pushed on their path as they leave.

In A Storage Ring, a beam of particles from a synchrotron is fed into a circular tube in opposite directions. The beams follow a winding path inside the storage ring and collide at certain places.

In a Cyclotron, (right) the particles travel between D-shaped plates. A magnetic field produced by magnets above and below the plates curves the path of the particles. As they speed up, they move outwards in a spiral. A strong electric charge kicks the beam from the cyclotron to hit the target.

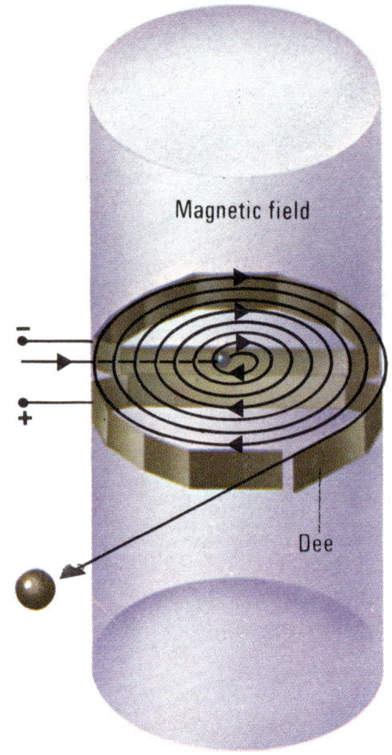

Into the Nucleus

Nuclear physicists are probing deep into the heart of the atom. Slowly the nucleus is yielding up its secrets.

Men have longed to find the material of which everything is made. They first found that all things are made of different atoms. Then they discovered that all atoms are made of three particles – electrons, protons and neutrons. When the neutron was discovered in 1932, scientists wondered if the nucleus contained any more particles. They also wondered whether the particles they knew were made up of even smaller particles. But the nucleus would not yield these secrets without a struggle. All kinds of nuclei had to be broken open and examined, and this takes a lot of energy. Ernest Rutherford had shown that bombarding nuclei with particles was the way to get into the nucleus. But more powerful particles were needed.

Nuclear physicists have therefore set about making machines to give particles more energy. They produce beams of particles and accelerate them to very high speeds. Then the beams are made to hit a *target* of a chosen substance. The nuclei in the target break down into other smaller nuclei, particles and rays. These can be seen and measured in special detectors. From their behavior, much has been learned about the nucleus and new particles have been discovered.

E. T. S. Walton

Sir John Cockcroft (1897–1967), a British physicist, and **Ernest Walton** (1903–), an Irish physicist, were the first men to change the nucleus with the aid of an accelerator. In 1932, they used a linear accelerator to bombard a target with a beam of protons. From their work have come many discoveries about what matter is made of.

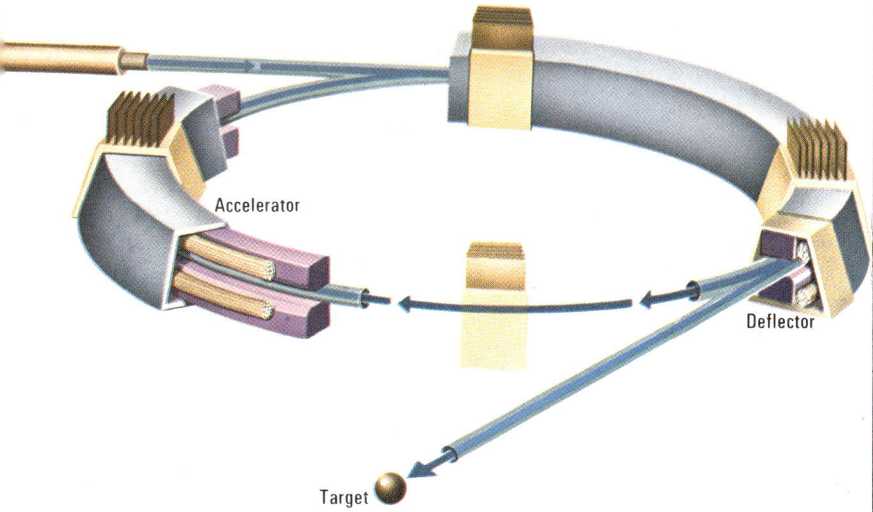

In a Synchrotron, the particles move in circles. The particles usually start from a linear accelerator. A beam of particles enters a circular tube surrounded by magnets. It travels around the machine many times, being accelerated by electric fields. Then the beam is deflected to strike a target.

In circular accelerators, the particles move in a spiral or a circle. In the *cyclotron*, particles are injected at the center and they travel outwards as they get faster, following a spiral path. At the outside, a deflector gives the particles an electric kick to make the beam leave the machine and hit the target. For very high speed beams, a *synchrotron* is used. In this machine, the beam travels in a circle, and the particles are kicked along by accelerators placed along the tube. When it reaches the required speed, the beam is shot out from the synchrotron to a target. The most powerful synchrotrons are huge. The largest is $1\frac{1}{4}$ miles in diameter.

To get collisions at higher speeds, nuclear physicists use *storage rings* – circular tubes in which beams can go round at high speeds for long times. By feeding two beams into a storage ring in opposite directions, the beams can be made to collide at certain points. The particles hit each other with immense force, making new nuclear reactions.

Particle Accelerators

Machines for producing high-speed beams of particles are called particle accelerators. They are sometimes also called atom smashers. But this is not a good name as they can build up atoms instead of smashing them. Accelerators use electric and magnetic fields to push and pull particles. Neutrons cannot be accelerated, because they have no electric charges. But electrons, protons, alpha particles and other small nuclei can be used. There are two basic kinds of accelerators: linear and circular.

In linear accelerators, the beam of particles moves in a straight line. The particles can be given a single boost by a strong electric field. But by giving them several boosts, they can be accelerated to very high speeds. The particles travel along a wide tube from which the air is removed. They go through a series of narrow *drift tubes*, which each "kick" the particles along with an electric charge.

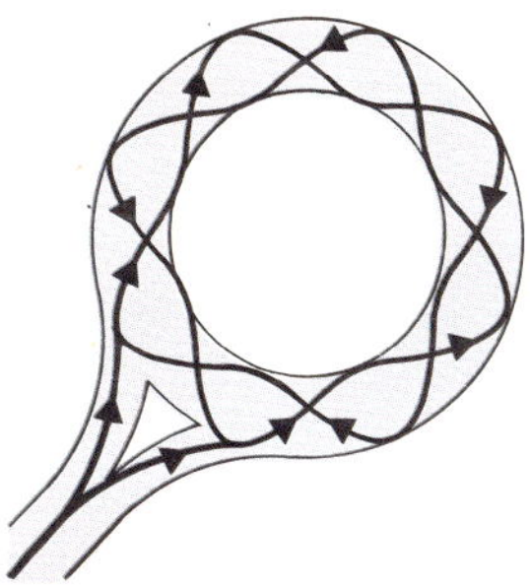

Above: Particle beams from a synchrotron can circulate along winding paths in opposite directions in a storage ring and meet head-on. Below: The beams can be stored separately in two rings to build up their speed before they meet.

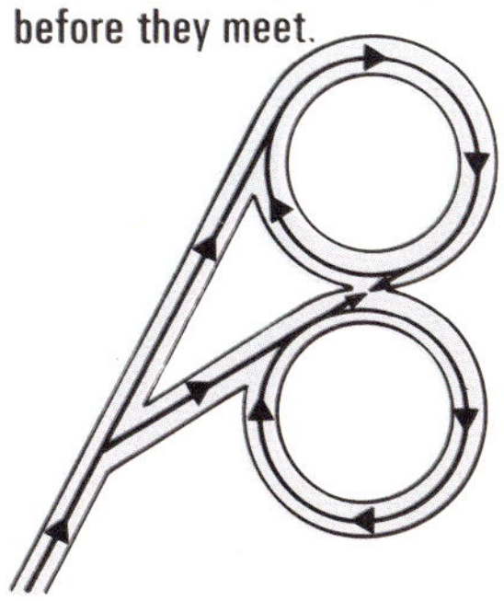

Below: Scientists at work on a proton synchrotron. Beams of fast-moving protons leave the accelerator along the narrow tube.

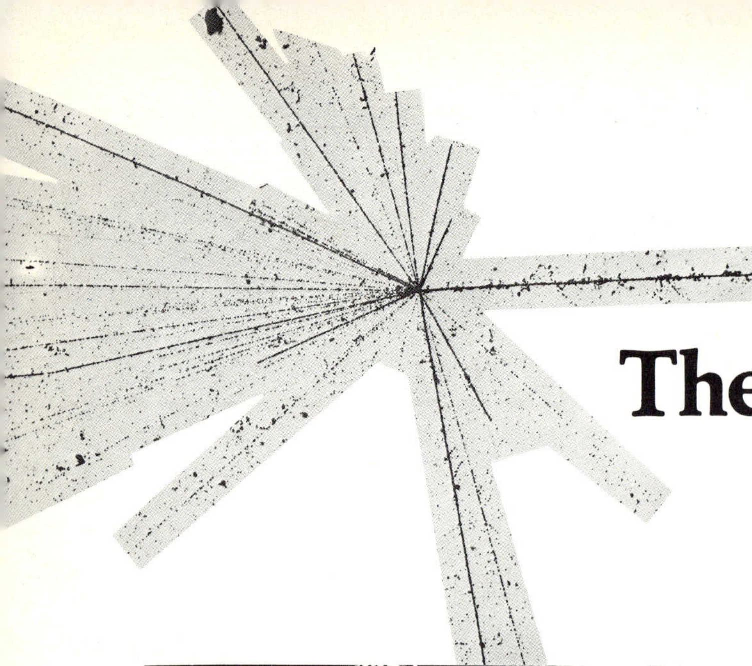

Left: Tracks in film show that a nucleus in a cosmic ray struck a nucleus in the film head-on. Both nuclei were shattered.

The Nature of Matter

In the fragments of the atom's nuclei, scientists have discovered a bewildering array of new particles.

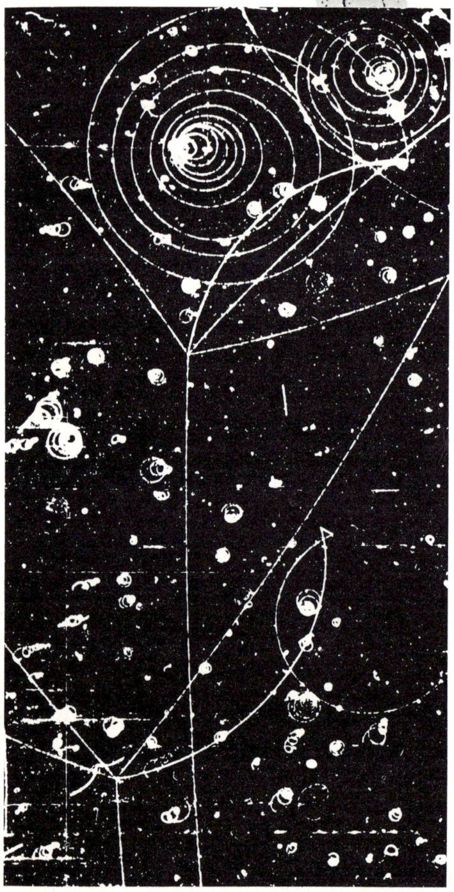

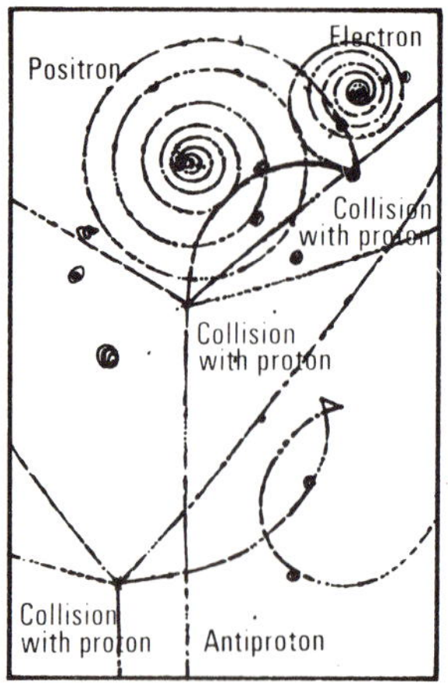

Above and left: Tracks in a liquid hydrogen bubble chamber show what happened when two antiprotons entered the chamber and collided with protons.

Below: Analysis of bubble chamber photographs is carried out with the aid of a computer.

Particle accelerators are not the only ways of obtaining high speed particles. Cosmic rays that strike the Earth's atmosphere from outer space consist of nuclei moving at very high speeds. They are so powerful that when a cosmic ray passes through an astronaut's head, he sees a brief flash of light. We do not see the flashes, which are harmless anyway, because the atmosphere prevents all but the most powerful nuclei from reaching the ground. To investigate cosmic rays, balloons are sent high in the atmosphere with packs of photographic film aboard. The packs are later retrieved and developed. Where a cosmic ray has passed, a track shows up. If it has struck the nucleus of an atom in the film, a *nuclear event* occurs. New particles are produced in the collision, and a star of tracks radiates out. In this way, a group of particles called *mesons* were discovered.

Fast particles from particle accelerators cause nuclear events to happen in the nuclei used as a target for the particle beams. The particles produced are detected in bubble chambers or spark chambers. A bubble chamber contains liquid hydrogen or a similar liquid just below its boiling point. When the particles pass through the liquid, it begins to boil. As it does so, a track of tiny bubbles is formed. A photograph is automatically taken of the tracks produced.

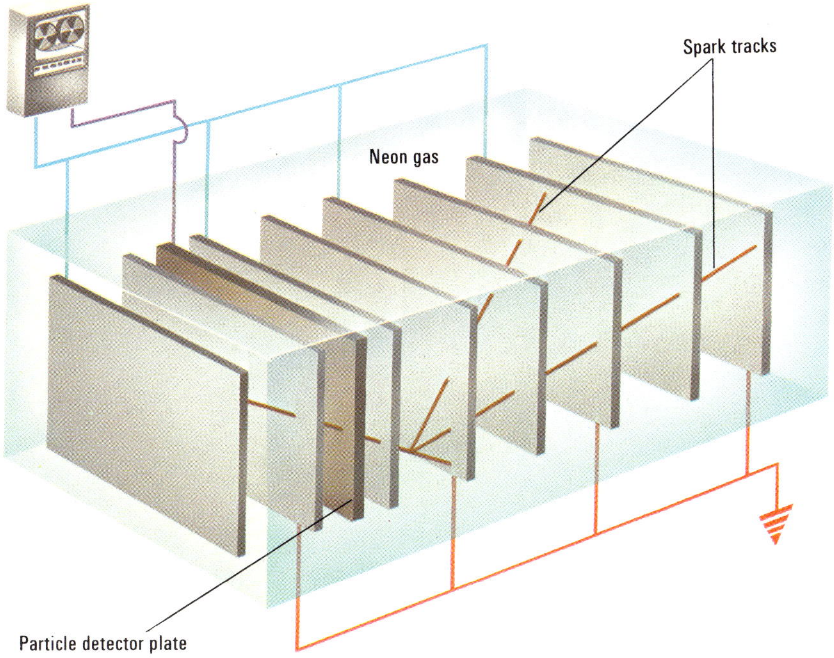

Above: A particle enters a spark chamber and leaves a trail of red sparks in the neon gas inside. The detector senses the particle's arrival and charges the metal plates.

Spark chambers contain a row of metal plates connected to a high voltage. When the particles enter the chamber, the high voltage is switched on by a detector in the chamber. The particles leave trails of sparks between the plates as they pass. The chamber contains neon gas and the sparks show up bright red like a neon sign. A photograph is automatically taken of the chamber.

Elementary Particles

From photographs of nuclear events, physicists have found that several new particles are produced by the collisions. One group of heavy particles are called the *hyperons*. The *leptons* are light particles, and include the *muon* and *neutrino* as well as the electron. In between these two groups come the *mesons*.

The hyperons, mesons, and the muon are all short-lived. Within a fraction of a second of being formed, they change into other particles. The neutrino is a very strange particle. It has no mass or electric charge, but it does possess spin. Neutrinos move very quickly, and can pass right through the Earth. The neutron is not stable when removed from the nucleus. It changes after nearly 17 minutes into a proton and electron and an *antineutrino*.

The antineutrino is one of a group of *antiparticles*. Every particle has an antiparticle in which the electric charge and other properties are opposite to those of the normal particle. In the case of the electron, for example, the *antielectron* has a positive charge instead of a negative charge. It is also known as the *positron*. Matter could exist that is made up completely of antiparticles. This matter is known as *antimatter*. For example, an atom of antihydrogen would consist of a positron orbiting an antiproton. When an antiparticle meets the corresponding normal particle, the two destroy each other and produce radiation. If antimatter were to meet normal matter, both would be destroyed. This could not happen on Earth, but some astronomers believe that it may be happening far away in outer space.

These various particles are the smallest units of matter that science has been able to detect. They are therefore known as *elementary* particles and also as *fundamental* particles. From them, everything in the universe is put together. Physicists cannot find out why there should be more than 40 kinds of particles. Further work is therefore going on to detect more particles and to find out if the particles we know are made of a few smaller, simpler particles. In this way, man is trying to discover exactly what he and the universe consist of.

PRINCIPAL ELEMENTARY PARTICLES
(excluding antiparticles)

name	charge	mass compared with electron	lifetime (seconds)
Proton	+	1836	stable
Neutron	none	1838	1010
HYPERONS			
Xi-zero	none	2570	$2{\cdot}9 \times 10^{-10}$
Xi-minus	–	2580	$1{\cdot}7 \times 10^{-10}$
Sigma-plus	+	2320	$8{\cdot}1 \times 10^{-11}$
Sigma-zero	none	2330	less than 10^{-14}
Sigma-minus	–	2340	$1{\cdot}66 \times 10^{-10}$
Lambda	none	2160	$2{\cdot}5 \times 10^{-10}$
Omega-minus	–	3280	$1{\cdot}1 \times 10^{-10}$
LEPTONS			
Electron	–	1	stable
Neutrino	none	0	stable
Muon	–	207	$2{\cdot}2 \times 10^{-6}$
MESONS			
Eta	none	1070	unknown
K1-zero	none	970	10^{-10}
K2-zero	none	970	$2{\cdot}8 \times 10^{-8}$
K-minus	–	960	$1{\cdot}2 \times 10^{-8}$
K-plus	+	960	$1{\cdot}2 \times 10^{-8}$
Pi-plus	+	273	$2{\cdot}6 \times 10^{-8}$
Pi-zero	none	264	10^{-16}
Pi-minus	–	273	$2{\cdot}6 \times 10^{-8}$

Facts and Figures

As the number of atomic power plants increases, man draws closer to a big new source of power for the future. Power from the atom will become vital as supplies of coal and oil dwindle.

The World's Nuclear Reactors

The chart on the left shows the number of reactors owned by the world's nations at the beginning of the 1970s, and those planned or under construction. Military reactors are not included.

The Rise In Nuclear Power

Electricity generated by nuclear power increased by more than ten times from 1960 to 1970, and is expected to rise another ten times between 1970 and 1980.

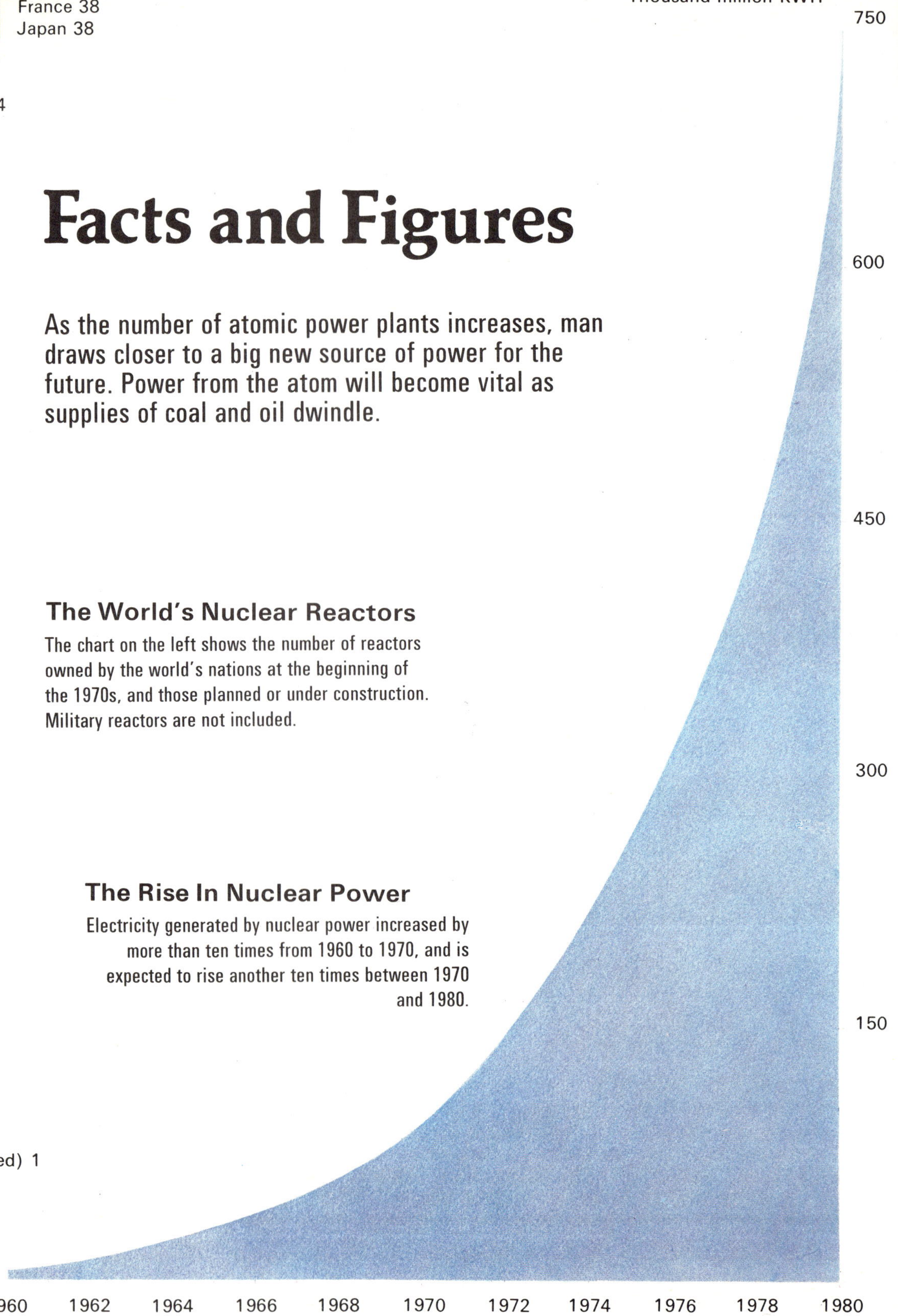

The World's Uranium Ore Deposits

The map shows the principal deposits that are being mined to make nuclear fuel. The United States has a fifth of the world's supply. There are large deposits of low-grade ore elsewhere.

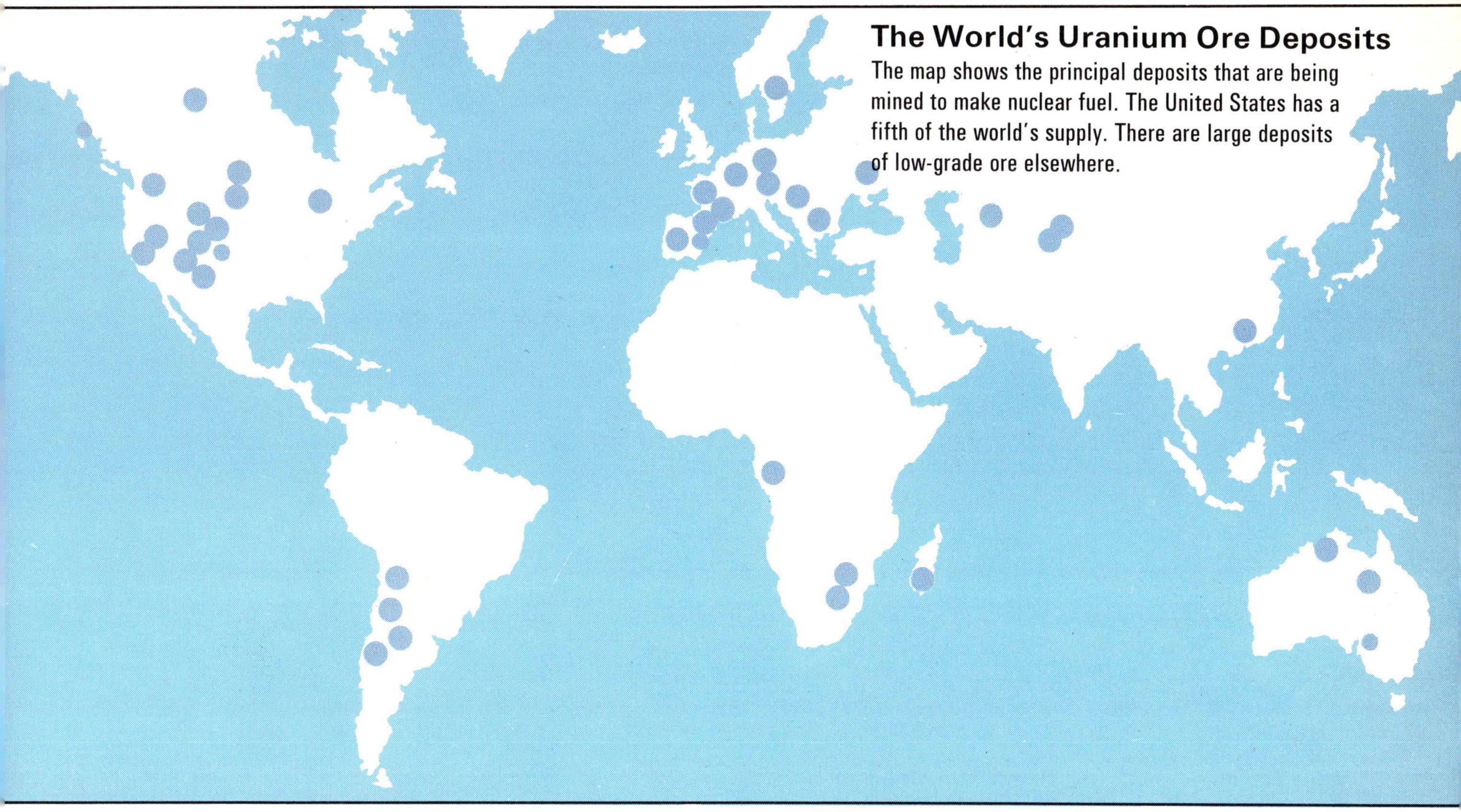

Nuclear Power Production

The chart shows the proportion of electricity generated by nuclear power in nuclear nations in 1970. The United Kingdom leads the world, but makes only a tenth of its electric power in this way.

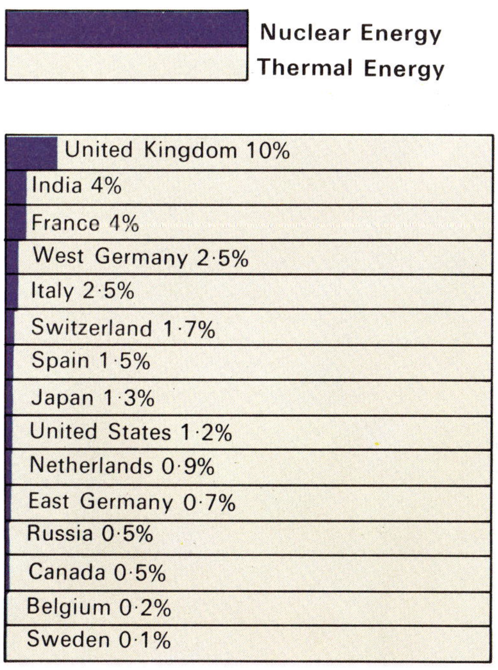

World Electricity Production

The chart shows that little more than 1 percent of the world's electricity was generated by nuclear power in 1970. Almost all electric power was made by burning fuel in thermal power stations or by hydroelectric stations.

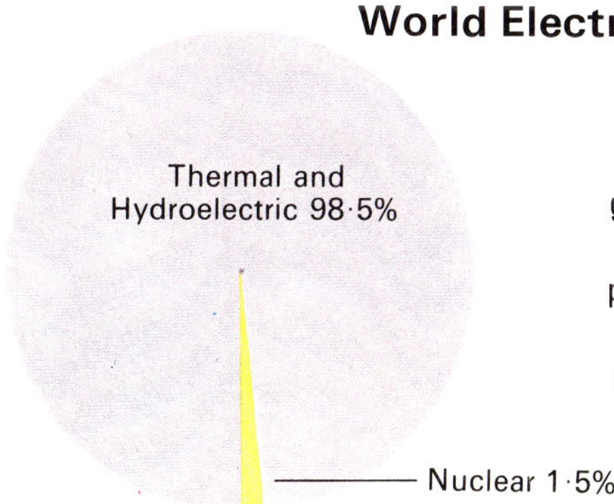

Nuclear Power Nation By Nation

The chart shows how the world's nations contributed to nuclear power production in 1970. The United Kingdom produced a third of the world's nuclear power, although it does not have the greatest number of reactors.

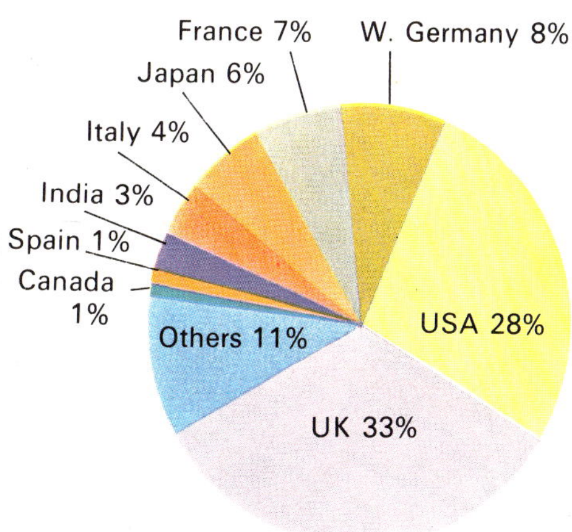

Index

Acknowledgments

Photographs: Page 2 Atomic Energy Authority; 3 RAE Farnborough; 4 University of Chicago; 6 J. Allen Cash; 8 USAEC; 11 Atomic Energy Authority (right), Birbeck College, Dept. Crystallography (center), Harvard University (bottom), Institute of Geological Sciences (top); 13 Mansell Collection; 14 Atomic Energy Authority; 15 Atomic Energy Authority (bottom), USAEC (top); 16 Atomic Energy Authority (bottom, right), Atomic Energy Authority (top), Saunders Roe Developments Ltd. (bottom left), US Atomic Energy Commission (top, left); 17 Atomic Energy Authority (bottom, right), Atomic Energy Authority (bottom, left), NASA (top, left), S. C. Johnson & Sons (top, right); 18 Atomic Energy Authority (center), US Atomic Energy Commission (top), Brookhaven National Laboratory (bottom); 19 Atomic Energy Authority (top left and bottom right), US Atomic Energy Commission (top, right), US Atomic Energy Commission (bottom, left); 20 Department of the Environment; 23 US Atomic Energy Commission; 24/25 Atomic Energy Authority; 25 United Press; 26 Atomic Energy Authority; 28 Atomic Energy Authority; 29 Novosti; 30 Bermuda Tourist Office; 31 California Institute of Technology and Carnegie Institution of Washington; 32 Atomic Energy Authority; 34 J. Allen Cash; 35 US Atomic Energy Commission; 37 US Atomic Energy Commission; 38 International Atomic Energy Agency (top), International Atomic Agency (left and bottom right); 39 CERN (top), ISPRA (bottom); 40 British Antarctic Survey; 41 C.I.C.; 42 CERN; 43 Atomic Energy Authority; 44 Professor P. H. Fowler, University of Bristol (top), US Atomic Energy Commission (center). Artwork: The Tudor Art Agency.

Picture Research: Penny Warn and **Jackie Newton.**